BIOETHICS for STUDENTS

HOW DO WE KNOW WHAT'S RIGHT?

General Editor
Steven G. Post
Case Western Reserve University

Carol Donley, Literature Consultant
Brian R. Frutig, Editorial Assistant

Macmillan Reference USA
Elly Dickason, Publisher
Hélène G. Potter, Editor
Anthony Coloneri, Editorial Assistant
Cynthia Crippen, Indexer

Editorial • Design • Production
Kirchoff/Wohlberg, Inc.

The entries in Bioethics for Students *are based, to a large extent, on the* Encyclopedia of Bioethics, *Revised Edition (1995). The Editorial Board of that publication consisted of the following:*

Editor in Chief

Warren T. Reich

Area Editors
Dan E. Beauchamp, State University of New York at Albany
Arthur L. Caplan, University of Pennsylvania
Christine K. Cassel, University of Chicago
James F. Childress, University of Virginia
Allen R. Dyer, East Tennessee State University
John C. Fletcher, University of Virginia
Stanley M. Hauerwas, Duke University
Albert R. Jonsen, University of Washington
Patricia A. King, Georgetown University
Loretta M. Kopelman, East Carolina University
Ruth B. Purtilo, Creighton University
Holmes Rolston III, Colorado State University
Robert M. Veatch, Georgetown University
Donald P. Warwick, Harvard University

BIOETHICS for STUDENTS

HOW DO WE KNOW WHAT'S RIGHT?

Edited by **Stephen G. Post**

Issues in
**Medicine,
Animal Rights,**
and
the Environment

VOLUME 2

Macmillan Reference USA
New York

Macmillan Reference USA
1633 Broadway
New York, NY 10019

Printed in the United States of America
Printing number
1 2 3 4 5 6 7 8 9 10

Bioethics for students : how do we know what's right? : issues in
    medicine, animal rights, and the environment/*Library of Congress
    Cataloging-in-Publication Data* edited by Stephen G. Post
        p. cm.
    Includes bibliographical references and index.
    ISBN 0-02-864936-2 (vol. 1 : alk. paper). -- ISBN 0-02-864937-0
(vol 2 : alk. paper). -- ISBN 0-02-864938-9 (vol. 3 : alk. paper).
-- ISBN 0-02-864939-7 (vol. 4 : alk. paper). -- ISBN 0-02-864940-0
(set : alk. paper)
    1. Medical ethics. 2. Bioethics. 3. Science--Moral and ethical
aspects.   I. Post, Stephen Garrard, 1951–   .
R724.B4825 1998
174' .2--dc21   98-29518
                                                          CIP

The paper meets the requirements of ANSI-NISO Z39.48-1992
(Permanence of Paper).

CONTENTS

Ethics and Law

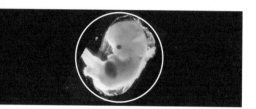

Fertility and Reproduction

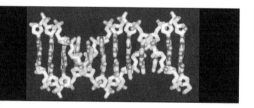

VOLUME 2

Genetics

Health Care

Mental Health

Population

Professional–Patient Issues

Stages of Life

Therapies

VOLUME 4

Therapies (Continued)

Transplants and Other Technical Devices

General Topics

Genetics

DNA TYPING

DNA typing is a scientific method for identifying people through the makeup of their genetic (hereditary) material. Typing involves comparing small portions of DNA from different sources and then determining whether these portions match. For example, samples for comparison would be obtained from a child and from a man who is believed to be the father. Or, a bloodstain and a sample from a criminal suspect might be compared to help prove the suspect's guilt or innocence. DNA typing can help identify people as accurately as fingerprinting. Furthermore, DNA typing can be used for a wider range of situations, since DNA is found in virtually all human cells. It can be taken from blood, **semen**, saliva, or hair roots, and it is highly resistant to decay and contamination.

semen the fluid that contains sperm

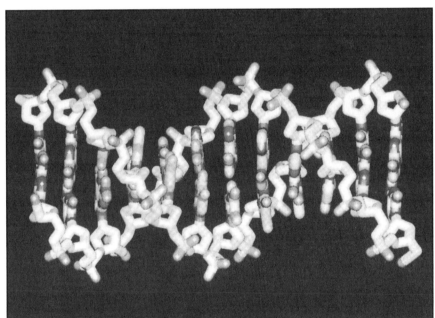

Computer graphic representing a segment of the molecule deoxyribonucleic acid (DNA).

DNA typing is a technology that can be very useful in many circumstances, such as in showing differences between serial crimes and copycat crimes, resolving disputes about parent–child relations, identifying remains, and helping parents locate missing children.

In theory, scientists could identify people with total certainty by examining their complete genetic sequence, which is a pattern of genes called a genome. There would be no error in identification since each person's genome is unique. But this would not be practical, because the human **genome** is spread over forty-six chromosomes, and each chromosome contains millions of molecules. DNA

genome all the genetic material of an organism; a complete set of chromosomes of an organism with the genes they contain

loci small regions of DNA sequences (singular, *locus*)

allele any of the alternative forms of a gene that may occur at a given locus

identification became possible with the discovery of gene **loci**, which vary in length and in other features from person to person. Using standard analysis techniques in the field of molecular genetics, investigators can compare these loci in DNA samples taken from different sources. But since the range of **alleles** found at any given locus is relatively small, several loci must be compared to make an accurate identification.

Accuracy of DNA Typing

The controversy over the accuracy of DNA typing focuses on the standards for declaring a match. The usual way to measure the length of DNA segments is to measure the extent of their "migration." This is a word for the motion of the DNA segment down a path through an instrument that contains an electronically charged gel. (The electric charge causes the segment to move through the gel.) But there is a problem with this technique. Because of a wide range of field (place where samples are obtained) and laboratory conditions, fragments of equal length may move at different rates. This phenomenon is known as band-shifting. There is also room for error in comparing DNA segments even without band-shifting. These problems have been dealt with partially by using uniform match standards and developing controls to reduce shifting. Also, the problem is helped by using new DNA-typing technologies that do not use electronically charged gel.

A second focus dealing with the accuracy of DNA typing involves the statistical interpretation of a loci match. Of concern here is the variation in the composition of the gene sequence. To calculate the probability that one person could be the only source of a sample being tested, investigators need to know how often matching alleles are found in the overall population. At present, genetic scientists sharply disagree about how much of a variation within matched alleles is meaningful. In other words, how does an allele variation in one group differ from such variations in other subgroups of the overall population?

Proper testing procedures are just as important to the accuracy of DNA typing as are correct interpretations of the statistics. DNA sample switching and contamination could be important sources of error along with misjudgment about band matches and allele variations. The results of special blind tests to see how well laboratories that carry out DNA typing are performing have not been satisfactory. Laboratory testing deficiencies point to the need for better quality control and for blind performance testing to be done regularly.

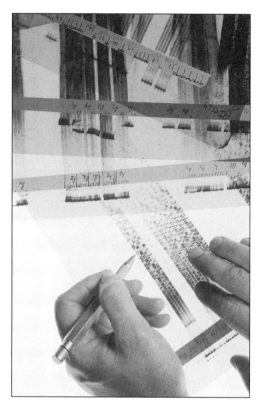

 The film shows dark patterns of DNA sequencing performed for genetic research.

see also

Genetics: Genome Mapping and Sequencing

Social Impact and Regulation

With the passage of state laws across the United States, DNA data banks have become widely permitted. Some states, as well as the FBI, store only DNA samples from convicted felons, especially sex offenders and others likely to repeat crimes. Other states, such as Louisiana, bank DNA samples from anyone arrested for a misdemeanor offense and retain the data even if the arrested individual is eventually proved innocent. In 1995, the FBI established its DNA Advisory Board to consider the many complex ethical issues of DNA banking, including individual privacy and control of genetic information. Because DNA is so accurate an identifying marker, its use in criminal justice systems is increasing around the world. The use of DNA samples in criminal trials, including samples taken from semen specimens in rape victims, is rising.

DNA and Privacy

The threat to an individual's privacy from DNA typing can come from two sources. First, samples for testing are easy to come by. Second, DNA typing results provide a great deal of information. Usable DNA material can be taken not only from blood or semen (which requires legal steps and a court order) but also from small amounts of tissue and saliva. The courts have not yet decided what restrictions should apply to obtaining such material or whether people have legally protected privacy interests in such material once it leaves their bodies.

The large DNA molecule itself is becoming a very rich source of personal information. Thousands of medical conditions have already been linked to certain regions of the genome. Additional important **genetic markers** are expected to be found for physical and psychiatric conditions. Most of the loci used in DNA typing, as described above, do not seem to be significant. However, the mapping of the human genome may find that certain loci are close to gene sequences related to medical conditions. For reasons of privacy, access to personal data could be limited by storing only one's DNA profile and not the material itself. But there are good reasons for preserving the DNA sample itself. For example, a new DNA analysis could be performed by an independent laboratory or even with a new technique. Yet it is important to have ways to prevent unauthorized use or misuse in **DNA data banks**. First, however, one must determine what exactly such an abuse would be.

DNA typing has been assessed by special panels of lawyers and scientists. The best-known panel is the National Research Council's commission on **forensic** DNA typing. Sometimes the people who serve on these panels have different opinions about the conflicts they are supposed to resolve. Nevertheless, the panels serve as models for evaluating other powerful, but possibly risky, technologies.

In the O. J. Simpson murder trial (1995–1996), many of the important defense arguments focused on the statistical reliability of matches between DNA samples taken from the crime scene and those taken from the defendant. Issues were also raised about the handling of these DNA samples.

genetic marker a gene or trait that serves to identify other genes or traits linked with it

DNA data bank a place where DNA information and material are stored

forensic relating to the application of scientific knowledge to legal problems

This entry consists of two articles explaining various aspects of this topic:

Historical Aspects
Ethical Issues

natural selection the theory that animals and plants best adapted to their environment tend to survive and that future generations show an increasing adaptation to that particular environment

artificial selection a deliberate breeding process of certain strains of plants and animals

At a 1912 National Conference of Charities and Corrections, a Miss Adams responded to eugenic theories as follows:

As an intelligent woman, but handicapped by blindness, . . . I would remind you that sixty per-cent of the blind sent out from schools are self-supporting. . . . When I observe the idle, selfish, shallow sons and daughters of the rich, spending their days in worth-less pursuits, making no contribu-tion of life and service for society, no answer to the great cry of hu-manity, I ask myself the ques-tion—who in the sight of God are the unfit?

HISTORICAL ASPECTS The word "eugenics" was first used in 1883 by the English scientist Francis Galton. Galton was a cousin of Charles Darwin and a pioneer in the use of mathematics to study biological inheritance. Galton took the word from a Greek root meaning "good in birth" or "noble in heredity." He wanted the term to mean the "science" of improving the human species by giv-ing the "more suitable races or strains of blood a better chance of prevailing speedily over the less suitable."

Darwin's theory of evolution taught that species changed as a result of **natural selection**. It was also well known that farmers and flower experts could produce permanent breeds of animals and plants with certain characteristics through **artificial selection**. Gal-ton believed that the human species could be improved in a similar way—that through eugenics, human beings could take charge of their own evolution.

"Positive" and "Negative" Eugenics

The idea of human biological improvement was slow to gain public support, but by the early 1900s, the eugenics movement had spread to many countries. Eugenicists everywhere shared Galton's under-standing that people might be improved in two complementary ways—by getting rid of the "undesirables" and by multiplying the "desirables." They spoke of "positive" and "negative" eugenics. Positive eugenics encouraged greater reproduction of people who were considered by eugenicists to be socially valuable. Negative eugenics encouraged those considered to be socially unfavorable to reproduce less or not at all.

Individuals with "good" genes were thought to be easily recog-nized by their intelligence, character, and physical appearance. Those with "bad" genes had to be weeded out. For the purpose of identifying such genes, in the early twentieth century eugenics paved the way to setting up the first programs of research in human heredity.

Social prejudices as well as socially positive goals shaped the eugenics movement. Eugenic studies claimed that the tendencies to commit crime, to engage in prostitution, and to have below-normal mental abilities were the results of bad genes. The studies also con-cluded that socially desirable traits were associated with the races of northern Europe, especially the Nordic race. Undesirable traits were identified with those of eastern and southern Europe.

Impact of Eugenics on Social Agenda

In practice, positive eugenics remained a theory only, but steps were being taken in the area of negative eugenics. Those who believed that **germ plasm** determines one's behavior insisted that "socially inadequate" people be discouraged or prevented from having children by urging or forcing them to undergo sterilization. Those holding this belief also argued for laws that restricted marriage for anyone they considered to be genetically undesirable and that kept such "undesirables" from immigrating to their countries.

In the United States, eugenicists helped pass the Immigration Act of 1924. This law greatly reduced eastern and southern European immigration to the United States. By the late 1920s, roughly half of all American states had passed eugenic sterilization laws. The laws were declared constitutional in the 1927 U.S. Supreme Court decision of *Buck* v. *Bell*, in which Justice Oliver Wendell Holmes delivered the opinion that "three generations of imbeciles are enough." The leading state in forced sterilizations was California, which as of 1933 had subjected more people to forced sterilization than had all the other states combined.

At the same time, a number of scientists and others were criticizing the teachings of eugenics. They argued that abnormal social behavior is above all the result of a social environment of poverty and illiteracy rather than of genes. They also argued that racial differences were not biological but cultural. These differences were the result of ethnicity rather than of germ plasm.

germ plasm the hereditary material of the germ cells; genes

▶ Officers examine immigrants at Ellis Island (1911). Eugenicists influenced the enactment of the Immigration Act of 1924, which, among other bans, restricted entrance to the United States of people with certain diseases.

The eugenics emphasis remained dominant in Germany, where eugenics reached its high point during the Nazi regime (1933–1945). Hundreds of thousands of people were sterilized for negative eugenic reasons, while millions of the "racially unfit" were sent to the gas chambers. Therefore, in the years after World War II, eugenics became very unpopular.

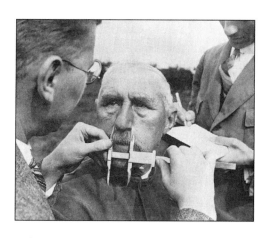

A Nazi official measures a German citizen's nose to determine his racial "purity." The Nuremberg Code's goal was to ban such practices as well as the types of medical experiments conducted in concentration camps.

Eugenics in Nazi Germany

1923 Adolf Hitler, while in prison, reads a eugenics textbook entitled *Outline of Human Genetics and Racial Hygiene* and incorporates eugenic ideas into his own book, *Mein Kampf*:

> The state must see to it that only the healthy beget children. . . . Here the state must act as the millennial future. . . . It must declare unfit for propagation all who are in any way visibly sick or who have inherited a disease and can therefore pass it on.

1923 Law passed requiring sterilization of people with schizophrenia, "feeble-mindedness," and others. By 1939, 350,000 people were sterilized.

1935 German law passed forbidding marriage and sexual contact between Jews and other Germans.

1937 Beginning of program to sterilize all German Gypsies, most of whom were murdered in the Nazi prison camps in 1943 and 1944.

1941 Beginning of plan to sterilize all one-quarter Jews and to kill all half and full Jews (the Holocaust).

Human Genetics: Eugenics Rid of Social Prejudices

Various British and American biologists tried to clean up the image of eugenics. They wanted to hold onto Galton's idea of human biological improvement while getting rid of the social prejudice that had invaded the practice of eugenics. They realized that a sound program of eugenics would have to be based on a solid science of human genetics. Human genetics needed to evaluate the respective roles of biology and environment, of nature and **nurture**, in the making of the human person.

The advances made in human genetics promoted the new field of genetic counseling, which gave future parents advice about what their risk might be of having a child with a genetic disorder. At first, genetic counseling could draw only on family histories and provide parents with the odds of conceiving a child with a **recessive** or dominant disease (or abnormality). Since the 1960s, genetic

nurture upbringing

recessive relating to one of a pair of genes that remains inactive; opposite of dominant

gene one unit in a chromosome that determines a trait

counseling has had the added help of tests that can identify whether a future parent actually carries an unhealthy **gene**. It can also determine prior to birth whether a fetus (unborn baby) suffers from genetic and chromosomal diseases or disorders.

Reproductive selection on a genetic basis—through the screening of parents, abortion of fetuses, or both—has been supported by liberal religious groups, nonreligious ethicists, and many feminists. Genetic selection has also raised fears among some members of minority groups and the disabled that it will lead to a revival of negative eugenics that may affect them most of all.

The Human Genome Project

These fears have been intensified by the Human Genome Project, the multinational effort, begun in the late 1980s, to obtain the sequence of all the DNA in the human genome. Once the complete sequence is obtained, it will be possible to identify individuals with unhealthy genes of a physical or antisocial type. The state may attempt to control reproductive behavior in order to discourage certain groups of people from transmitting unhealthy genes throughout the larger population.

Genetics: Genome Mapping and Sequencing

Economics may become a powerful motive in setting up a new negative eugenics program. In the United States, as health care becomes more and more the responsibility of taxpaying citizens, there is a greater possibility that taxpayers will oppose paying for the care of those whose genetic makeup dooms them to severe disease or disability. Even in countries with national health systems, public officials may feel pressure to encourage, or even to force, people not to bring genetically diseased children into the world. This pressure would be exerted to keep public health costs down.

A number of factors are likely to outweigh a widespread revival of negative eugenics. For instance, eugenics programs tend to be put in place under dictatorships. Although democratic governments may not have been strong enough to resist completely the violations of civil liberties connected with the early eugenics movement, they did fight against them effectively in many places. The British government refused to pass eugenic sterilization laws. So did many American states. Where such laws were passed, they were often unenforced. The cruelties of state-sponsored eugenics in the past have turned most geneticists and the general public against such programs. In addition, people with handicaps or diseases, as well as minority groups, have more political power than they had in the early twentieth century.

Human Genetics in a Market Economy

The progress in human genetics and biological science has created the potential for a kind of individual eugenics—families deciding

what kinds of children they wish to have. Although most parents would now probably prefer just a healthy baby, in the future they might be tempted by the opportunity to have "improved" babies, children who are likely to be more intelligent or more athletic or better looking (whatever such terms might mean).

Despite the shadow of eugenics associated with the Human Genome Project, many observers believe that its ethical challenges lie not in attempts at eugenic policies so much as in the enormous amount of genetic information that the project will produce. These challenges center on how that information will be controlled, distributed, and used within a market economy.

Many critics have pointed out that the flood of new human genetic information will present challenges to social fairness and equal treatment. They have warned that employers may try to deny jobs to applicants with a low resistance—or a supposed low resistance—to disorders such as manic depression or illnesses connected to conditions of the workplace.

Many analysts concerned about such possibilities have argued that an individual's genome information should be protected as a privacy issue. Legal and insurance analysts have pointed out that insurance premiums depend on judging the risk factors of each potential client. If a client with a high genetic medical risk is not charged a high premium, then that person receives a high payout at low cost to him- or herself but at high cost to the company and to low-risk policyholders. Insisting on a right to privacy regarding genetic information could sometimes lead to unfair consequences—at least under the largely private system of insurance in the United States. The Human Genome Project is less likely to encourage a new eugenics than it is to present disturbing challenges to public policy and private practices as to how human genetic information will be controlled and used.

ETHICAL ISSUES There is much confusion with regard to suitable definitions used in discussing eugenics. Words like "normal," "harmful," "desirable," "undesirable," and "improvement" are scientifically unclear. An exact definition of a "harmful **genotype**" remains unclear, in spite of the increased understanding of the causes of genetic disease.

Does a person who carries one or more recessive genes that can cause a genetic disease such as sickle-cell anemia have a harmful genotype? Being a **carrier** often means having a stronger resistance to disease rather than suffering from the disability hinted at by a "harmful genotype." A critical issue in genetics is whether it is morally acceptable to move from the "is" of this kind of naturally occurring human genetic variation to the "ought" of actually using such a variation (for instance, through genetic selection) to bring about eugenic goals.

genotype the genetic makeup of an individual or group

carrier someone who "carries" the gene for a genetic disorder

The morality of any eugenic strategy depends on the definition, human cost, and justification for using genetics to achieve human "improvement." A naturalistic view in ethics holds that the place of humans in the order of nature makes it clear that evolution has a direction. From this we can infer that improvement is necessary, because it is an extension of the natural order. Some evolutionary biologists argue against this view, pointing out that nothing in the evolutionary program suggests an inborn tendency toward "progress."

Because eugenic methods have been suggested throughout history by those who are "best off" in terms of political power, eugenics raises issues of justice and fairness. Would some people, through no fault of their own, benefit less than others from genetic research and eugenics?

Eugenics provides a lens through which to see forms of reproductive **coercion** Ethicists have long questioned the use of coercion through government policy or social persuasion to influence the reproductive choices in certain groups.

coercion the process of bringing something about by force or threat

Issues Affecting Women

Since many eugenic policies require the cooperation of women, such policies can discriminate by causing unequal physical, psychological, and social risks to women. Examples include female-centered **sterilization** programs and birth-control policies, and the use of prenatal (before birth) diagnosis and abortion.

sterilization the act or procedure of making a person incapable of sexual reproduction

The use of genetic diagnoses to insist on the testing of **embryos** also has a direct impact on women's reproductive freedom. The promotion of genetic screening and pre-embryo testing will put social and psychological pressure on women to choose "quality" in the children they bear over simple reproductive success. The venue for such genetic selection would most likely be a genetic counseling clinic.

embryo a developing unborn baby (from fertilization to about eight weeks)

Genetic Counseling and Eugenics

Eugenic considerations are not usually identified openly in programs that might influence reproductive decision making. Most genetic counselors recognize that many existing genetic programs, including prenatal diagnosis, newborn screening, and carrier screening may have indirect or serious eugenic or **dysgenic** effects. Genetic counselors try to stay neutral when it comes to attaching values to each possible decision. But the information they share with their clients may be dysgenic, eugenic, or neutral, depending on the circumstances.

dysgenic promoting survival of the weak or diseased at the expense of the strong and healthy

A closer look at the counseling process and the information shared shows that eugenic aims may be incorporated in a hidden way into the options and diseases chosen as "suitable" for selective abortion. An example would be sex chromosome abnormalities that

are usually linked only with physical and mental abnormalities. Revealing to anxious parents the genetic condition of a fetus with Turner's syndrome (where the absence of the X chromosome leads to shortness, webbed neck, and minor nervous-system deficits in girls) is just such a situation.

The ethic of value-free counseling leads to an approach not to interfere with clients' rights. This may allow for dysgenic outcomes. The process of aborting a fetus with a harmful recessive disease and then compensating for the loss of the expected child by trying to have more children ultimately results in a slight *increase* in the number of times the recessive gene shows up over many generations. This is true because such reproductive compensation causes two-thirds of all of the live children to be carriers of the recessive gene in question, instead of the one-half normally expected without prenatal diagnosis.

Justifying Eugenic Policies

The appropriateness of providing state-sponsored services to encourage reproductive choices is an ethical issue that is less obvious than forced sterilization, required genetic testing, or state-encouraged selective abortion. The major ethical questions raised by intentionally adopting (or ignoring) possible eugenic or dysgenic outcomes are based on the following considerations, among others: (1) our duties to present versus future generations; (2) the obligation not to do harm versus the duty to do good; and (3) the obligation of health professionals to provide for the needs of individuals or families versus that of protecting the **gene pool** as a whole.

From a purely abstract point of view, it is possible to consider the idea of eugenics with the assumption that there is nothing wrong in itself with improving the human condition. It is not wrong as long as society's values are strengthened and obvious harms to unprotected populations are avoided.

Validating Genetic Policies

Because genetic disease includes suffering, and the lessening of suffering is a widely recognized goal of medicine, all potentially eugenic policies that affect a whole population can be measured against this claim for endorsing genetic change. It is reasonable to ask if adopting a eugenic policy would significantly reduce human suffering in the long run. Where and when does the genetic standing of a given group of family members justify genetic intervention?

While eugenic interventions are ethically questionable because they often include coercion, there are special situations when such policies may be acceptable. Among the conditions that need to be met are at least the following:

Fertility and Reproduction: Abortion

By 1920, twenty-five states had laws requiring sterilization of many criminals and others considered mentally unfit. Margaret Sanger proclaimed: "Breed more children from the fit, less from the unfit." President Theodore Roosevelt strongly supported eugenic laws. Eugenic theories dominated in many American scientific circles. By 1930, an estimated fifteen thousand people, mainly those with varying degrees of mental retardation or illness, had been sterilized without consent.

gene pool the collection of genes of all the individuals of a particular population

1. The genetic condition of a particular group has been shown to be sufficiently risky to justify an intervention.
2. The individuals who will be affected by the policy are given a voice in the decision making.
3. The results sought are justified by the group's own standards, at costs the group finds acceptable.
4. The means are necessary and ethically acceptable to the affected individuals.
5. Policymakers can give reasonable assurances that both the risks and the benefits of the proposed policy will be equally and justly distributed.
6. Society's goals can be accomplished without intruding on other cultural or basic values.
7. The eugenic policy represents the least forcible option to obtain the agreed-upon purpose.

Ethics Versus Politics

Constitutional and ethical arguments against eugenics are based on the importance of privacy in reproductive decision making. Any moral excuse for eugenics must prove the lawfulness of the state's right to limit fundamental reproductive freedoms based on individual differences.

When policy decisions restrict the reproductive options of certain groups, they provide a kind of back door to eugenics. Such

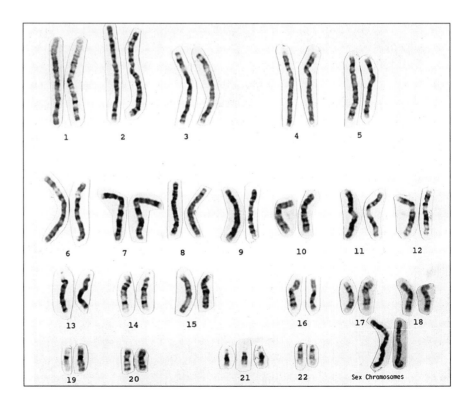

▶ A karyotype of a girl with Down syndrome (chromosome 21 trisome). "Negative" eugenics held that individuals with such a condition are socially undesirable.

policies can lead to a social view that lowers the worth of groups with certain genotypes. Policies of reproductive screening, reproductive exclusion, or both can be mistakenly connected with "solutions" to society's problems. Such events as testing of newborns for an extra Y chromosome (mistakenly believed to be associated with criminality) are burdened with moral importance. This is especially true when socially identified or labeled genes are found in certain ethnic or cultural groups with a long tradition of suffering social discrimination. Such genetic lines will become more evident as data are produced to map and sequence the human genome. Such data raise questions about the possible misuse of genetics in legal activities, especially where including ethnic differences may be crucial for accurate testing.

The Human Genome Program

As the net of genes identified through the Human Genome Program continues to expand, the chances of applying eugenic ideas in setting diagnostic policies will certainly increase. This project will provide a database for genetic screening and testing that could give couples greatly improved information about the genetic results of their reproduction and thus the potential to select sperm, eggs, or both with certain genotypes before **fertilization** occurs.

The Human Genome Program would also identify groups of individuals whose number of unfavorable genetic **mutations** is unusually high compared with that of the surrounding population. Such groups probably exist generally, and high concentrations may be found among offspring whose parents were exposed to certain chemicals or to radiation.

Such an outcome suggests that ethical considerations of the impact of eugenics on reproductive choices need to play an essential part in assuring equal justice for all. Deciding which groups benefit by virtue of their good "gene quality" and which groups may be hurt by the continued neglect of their "genetic burden" will continue to present ethical dilemmas in the future. With so much genetic knowledge available, ignoring genetic differences is itself a policy that requires justification.

fertilization in humans, the union of an egg and a sperm

mutation a significant and basic change in living organisms

Genetics: Genome Mapping and Sequencing

Related Literature

Aldous Huxley, in *Brave New World* (1932), imagines what the world might be like 632 years After Ford (dating from the invention of the automobile). In this society, all reproduction is controlled by the state. Human eggs fertilized in petri dishes are placed in different solutions, which enhance or arrest their growth according to whether they are designed to be Alpha Plus

Intellectuals or Epsilon Minus Morons. All "births" take place in the Central London Hatchery and Conditioning Center, where humans are mass produced according to desired characteristics determined by the state and where babies are conditioned after their birth to fit into the class for which they were designed. Although this novel was written in 1932, it anticipated some of the issues raised in modern reproductive technologies, from the devaluation of the individual and the family to the power of technologies to coerce behavior.

GENETIC COUNSELING

In 1953, James Watson constructed a cardboard model of the gene, fitting together complementary nucleotides (adenine, guanine, thymine, and cytosine). Watson and Francis Crick suggested a three-dimensional model known as the "double helix," in which two ladders of nucleotide pairs are twisted about each other. In 1962, the two scientists were awarded the Nobel Prize in medicine for their work.

Genetic counseling may be described as the communication about genetic concerns between a health-care provider and a patient or family member. These concerns may be about the birth of a child, medical problems, reproductive testing choices, a family history of bad health, or the diagnosis of an inherited condition. There are over four thousand genetic conditions, chromosome disorders, and birth defects that can result in miscarriage, stillbirth, early death, or problems in childhood or adulthood. Those who ask for this type of counseling service have questions about why a condition occurred, what the chances are that it may happen again in the future, and how they can cope with the results of a diagnosis. These services are often provided by a team of clinical genetics specialists in a medical genetics clinic within a hospital or in clinical outreach settings or professional offices. The medical, educational, and emotional needs of the patients and their family members should all be taken into account in the counseling sessions.

Genetics Evaluation and Counseling Services

A correct diagnosis is necessary in order to care for patients' needs. This is true even if the person asking for counseling does not have a genetic condition. The diagnosis of the patient or relatives determines the inheritance pattern in the family and the risk information that is provided. The diagnosis may be determined from family history information, pregnancy history, medical records, clinical examinations, or test results. The physician geneticist on the team usually performs this part of genetic counseling. Medical and **pediatric** geneticists are doctors who have completed special medical training in clinical genetics. They are able to recognize rare patterns of physical features unique to particular genetic disorders, interpret test results, and recommend ways to deal with related medical concerns.

pediatric relating to a branch of medicine that deals with the development, care, and diseases of children

▶ During a genetic counseling session, a doctor discusses the couple's genetic information contained in the counselor's binder.

Genetics: Eugenics

psychotherapeutic used in the treatment of mental or emotional disorders

Standard of Practice

It is important to stress the use of the term "counseling" in referring to this clinical genetics practice, in order to set it apart from giving directive genetic advice (eugenics). A patient's freedom to make his or her own decisions should be respected and upheld, particularly when it comes to reproduction. A definition of genetic counseling, published in 1975 by the American Society of Human Genetics Ad Hoc Committee on Genetic Counseling, stressed the educational part of the service. This definition indicates that the most important aspect of genetic counseling is its nondirective approach. A very important principle in genetic counseling is the patient's freedom to choose a course of action.

Nondirectiveness. The term "nondirectiveness" was widely used in the 1970s. Although the use and meaning of the term varies greatly, "nondirective" was borrowed from the writings of the psychologist Carl Rogers to describe a non-advice-giving counseling process. Rogers first used the term in 1942 to describe his **psychotherapeutic** approach of not advising, interpreting, or guiding his clients. The concept caused confusion as Rogers realized that his very presence in a relationship with a patient had many "directive" sides to it. By 1978, he had begun using the term "person-centered" to better describe his therapeutic approach. Clinical geneticists and genetic counselors often use the term to describe different aspects of their practice, but there have been arguments about whether genetic counseling can be completely nondirective.

Making recommendations. Nondirectiveness does not mean that genetic counseling is without values. However, geneticists understand that their own values and ideas often are not the same as those of their patients. Rather, geneticists and counselors are very clear about certain values, such as the patient's role in making his or

As opposed to the destructive history of eugenics prior to the end of World War II (1945), in which sterilization and even the killing of the "unfit" were tolerated, genetic counselors and modern society emphasize respect for reproductive freedom. Individuals who are tested for genetic mutations make their own choices about reproduction, without any interference from counselors, society, or government

informed consent consent to medical treatment by a patient—or to participation in a medical experiment by a subject—after achieving an understanding of what is involved

Ethics and Law: Autonomy

her own decisions, especially about childbearing. It is impossible for geneticists and counselors to know what they would do if they were in another person's situation.

Genetic counseling clearly has directive parts to it. The education of patients is directive in that the counselor shares certain information with them. Since they often come asking for such information, this activity is unavoidably directed. Managing medical help and making referrals to support groups or agencies are also directive. Refusing to offer a carrier test to a minor or to offer prenatal testing for selecting the sex of a child are other examples of directed activities. Genetic counseling may even involve making recommendations.

Goals of Counseling

The main goal of genetic counseling is to reduce emotional suffering related to genetic conditions and to try to answer patients' concerns. Clinical genetics professionals who think they know better than their patients what are "correct" or "good" decisions are not providing patient-centered care. Education and counseling *before* testing are important parts of any genetic-testing or population-screening program. If this important practice is not observed, clinical genetics professionals risk returning to the tragedies of eugenic programs.

Ethical Issues

Genetic counseling should be guided by several ethical principles and human values that are judged by most workers in the field to be of utmost importance. These include the following: (1) patient freedom; (2) confidentiality; (3) accuracy and truth-telling; and (4) **informed consent**. It is also necessary that cultural and ethnic factors be taken into account.

Patient freedom. If patients are to make independent decisions, they must be fully informed about the disorder in question. They must be free of pressure and aware of all the possible choices, and have access to facilities and services to carry out their decision. In its purest sense, the nature of the decision is not an issue as long as the patient has decided that the decision is in her or his best interest. In this model of counseling, the counselor makes every effort to not give any suggestion, directly or indirectly, to the patients as to what decision they should make.

Confidentiality. All information regarding the patient's case is covered by the guarantee of privacy and confidentiality that is required of health professionals. The medical geneticist or genetic counselor should get permission from the patient to contact other family members to inform them that they are at risk for a serious genetic disorder. Usually, this is not a problem. Most clients quickly

consent to having their relatives contacted or are willing to contact them themselves. But in at least two situations the genetic counselor may face an ethical dilemma concerning the giving of information to relatives:

- The condition is *not* treatable but can be diagnosed during a pregnancy, so a couple at risk could decide to terminate the pregnancy. Or individuals at risk might wish to take special tests that could predict a genetic disorder and use this information to change their life plans.

- The condition *is* treatable and can be cured, or the symptoms can be greatly reduced by safe and available therapy, or the expression of the disorder can be prevented if it is found before the symptoms appear.

The obligation to keep patient records and genetic information strictly confidential may not be kept when there is sufficient reason. This may be done only if there are special circumstances.

Accuracy and truth-telling. A major part of the genetic counseling process is the information given by the patient about his or her medical and family history and the complete genetic and medical information about the disease provided by the counselor. The patient needs accurate information, including the correct diagnosis, in order to choose the best course of action. Truth-telling is an essential ingredient of the relationship between genetic counselors and their patients. Part of the trust that exists between them is based on this virtue. The genetic counselor should provide truthful, accurate, and complete information to the patient about the genetic disorder being discussed.

Ethics and Law: Confidentiality;
Professional–Patient Issues:
Professional–Patient Relationship

A doctor talks to a pregnant woman about genetic screening for Down syndrome.

Uses of genetic testing:

1. Diagnostic genetic tests can confirm the diagnosis of certain diseases in patients.

2. Predictive genetic tests can indicate whether a person who at present is free of a disease will eventually develop that disease because of a dominant mutation.

3. Susceptibility genetic testing can determine whether a person who is at present free of a disease is at some greater risk for that disease than is usually the case.

4. Prenatal genetic testing can determine whether the fetus carries any gene mutations that will result in disease. The mother may or may not decide on an abortion, which is her choice alone.

Informed consent. Since a major part of genetic counseling is the sharing of information, and since the patient is encouraged to make his or her own decision, problems or conflicts with informed consent are rare. Informed consent is especially important in the counseling process when a procedure may result in harmful or uncertain outcomes.

Ethics and Law: Information Disclosure, Truth-Telling, and Informed Consent

GENETIC ENGINEERING

In April 1988, the U.S. Patent Office announced that it "now considers non-naturally occurring nonhuman multicellular living organisms, including animals, to be patentable." One year later, the first animal patent was issued to Harvard University for a mouse that was genetically engineered to have very little resistance to developing tumors, making it useful for cancer research. The Du Pont Corporation then produced the patented mouse in large numbers to be sold to businesses.

Animals and Plants

The patenting of plants and microorganisms had not caused a problem for the public, but the patenting of animals did. One of the concerns was that the issue of patenting animals was too complex to be decided only by the Patent Office. Instead, critics said, the issue should be settled as a matter of public policy by the U.S. Congress. Exactly what were the moral and lawful issues was not clear, and people found it very difficult to separate moral issues from emotional ones.

Research: Patenting Organisms

Some of the strongest criticism of the genetic engineering of animals can be summed up in this statement based on the Frankenstein story: "There are certain things humans are not meant to know,

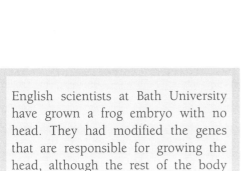

Corn tissues at various stages of growth, Sungene Technologies Laboratory, Palo Alto, California.

English scientists at Bath University have grown a frog embryo with no head. They had modified the genes that are responsible for growing the head, although the rest of the body was still intact.

do, or meddle with." Such a view is centered on the belief that genetic engineering of animals is wrong even if it produces important benefits and little or no harm. This position is mostly a religious viewpoint. For example, a statement signed by twenty-four religious leaders from a variety of faiths claims that "the gift of life from God, in all its forms and species, should not be regarded solely as if it were a chemical product subject to genetic alteration and patentable for economic benefit."

Along the same lines, one can find arguments against placing human genetic material in animals. One example of this is the case of the human growth hormone gene that has been inserted into pigs to produce larger animals. The statement noted above from the religious leaders says that "the combining of human genetic traits with animals . . . raises unique moral, ethical, and theological questions, such as the sanctity of human worth."

Another possible ethical problem concerns keeping the "species integrity" intact. Critics claim that making major changes to a species by humans is in itself wrong. They also state that species that are unchanging are the building blocks of nature. All of these points require much support, because modern biology rejects the unchangeable belief about species, and humans have been changing plant and animal species significantly for much of history. If genetic engineering is consistent with what humans have always been doing, then the burden is on the critic to show why bringing about the same result through a different technology represents a moral problem.

The Risks of Genetic Engineering of Animals and Plants

Possible dangers connected with the genetic engineering of animals obviously come from the speed in which major changes in

organisms can be achieved. Traditional "genetic engineering" was done by selective breeding over long periods of time. This allowed plenty of opportunity to observe the unfavorable effects. With newer techniques, however, scientists are doing their selection "in the fast lane."

Weakening of the species. First, unexpected results may affect the organism that is being rapidly changed. For example, when wheat was genetically engineered to resist blight, the wheat's backup gene for general resistance to disease was ignored. As a result of this oversight, the new organism was open to all sorts of viruses that destroyed the crop in one generation.

Harmful effects on humans. Second, the particular characteristic engineered into the organism may have unexpected harmful effects on humans who interact with it. This could happen to people who consume the new life-form as food. It might be possible to genetically engineer faster growth in beef cattle in a way that would increase certain levels of hormones that turn out to cause cancer in human beings.

Multiplying problems and decreasing diversity. A third set of risks copies and increases problems built into selection by breeding. These include narrowing the **gene pool**, the tendency toward genetic sameness, the rise of harmful recessive gene traits, the loss of hybrid energy, and, of course, the greater sensitivity of organisms to destruction by microorganisms that cause disease, as has been shown in some genetically engineered crops.

A fourth group of risks comes from changing the disease-causing microorganisms that live in or on the animals being altered. This can happen in two ways. First, one could unknowingly create an environment that could produce a **mutation** of that disease-causing microbe to which the altered animal would not be resistant. These

gene pool the collection of genes of all the individuals in a particular population

mutation a significant and basic change in living organisms

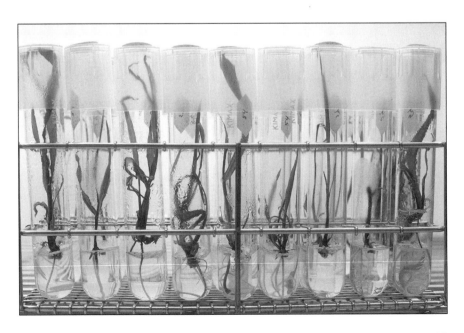

Genetically engineered corn plants grow in controlled conditions for desired genetic traits at a Sungene Technologies Laboratory at Palo Alto, California.

new organisms then could be infectious to these or other animals, or to humans. Second, one could inadvertently change the microbes by changing the environment in which they live. This could make them dangerous to humans or other animals. The more abrupt the change, the more difficult it is to estimate the effects on the microbes.

Limiting the Risks of Genetic Engineering

Supporters of genetic engineering of animals respond to the list of risks by suggesting ways in which these risks could be kept to a minimum. They include the following: (1) keeping genetically engineered farm and laboratory animals away from all other animals; (2) strictly limiting the release of such nondomestic animals as fish; and (3) engineering such traits into the animals as nutritional needs that can be met only by human-supplied diets, so that the animals could not survive in the wild. Supporters also describe many possible benefits of genetic engineering. Genetically engineered animals could increase productivity of food and fiber. The "super pig" and "super chicken," which are much larger than normal animals, are two examples. Developing such animals could help the economic growth of the country that produces them. Genetic engineering could lower environmental damage by fitting animals into environments that were previously nonproductive; and through what has been called "molecular pharming," mammals can be genetically engineered to produce in their milk drugs and chemicals that are rare or expensive. Scientists have already developed goats that produce milk containing TPA, a drug used to dissolve blood clots. Supporters also stress the value of genetically engineered animals to biomedicine. Such animals can, in theory, model a variety of diseases and conditions. Most notably, they can be used as models of human genetic diseases, which are some of the most devastating and untreatable diseases. In addition, animals can be used to study the safety and effectiveness of gene therapy to treat these diseases.

Should Genetic Engineering of Animals Be Banned?

Supporters of genetic engineering of animals also argue that it is simply not possible to ban it. Unlike other possibly dangerous technologies, such as nuclear technology, genetic engineering can be done cheaply and easily. A ban in a particular country would simply send the technology to places without a ban, and would get in the way of the development of international regulations. It could also harm a country's ability to compete economically with the other countries of the world.

An embryo-transfer specialist holds a calf born from an artificially inseminated embryo carried by a surrogate mother.

▶ A large transgenic salmon is contrasted to two nontransgenic siblings alongside a ruler. The transgenic salmon has had genetic material from another species added to its genome.

artificial selection a deliberate breeding process of certain strains of plants and animals

▲ The Austrian geneticist Gregor Mendel (1822–1884) experimented with crossbreeding of garden peas, which enabled him to formulate the principle of heredity.

A complete ban on genetic engineering of animals does not appear to be practical. Critics concerned about the ethical issues are hoping for a long-term suspension of animal patenting while society debates all the issues involved. Whether or not such regulation is put into law, a debate is necessary to provide workable regulations for controlling the risks to human beings and the environment as well as the animal suffering that goes along with the development of this powerful new technology.

The genetic engineering of plants has caused far less disagreement in society. Humans are used to tampering with plants to create new species. The tangelo, for example, is a cross between a tangerine, a Mandarin orange, and a grapefruit. It is estimated that 70 percent of grasses and 40 percent of flowering plants represent new species created by humans through **artificial selection**. However, putting animal genes into plants, as has been done by putting flounder genes into tomatoes to prevent freezing, arouses a negative reaction from many people in society. This reaction seems to be a fear of the dangers that might result from eating the product.

So, when genetic engineering of plants is compared to genetic engineering of animals, one sees different reactions. There is very little emphasis on the natural wrongness of plant engineering, and almost no concern about plant welfare, since few people believe that plants are capable of feeling. Most social concern comes from possibly disturbing results of such engineering on humans and the environment. The environment, especially, represents a legitimate concern, since there are many existing examples of plants being placed in unintended environments with harmful results.

A medical researcher conducts an experiment in gene therapy. Gene therapy is now being applied to treat cystic fibrosis.

Human Genetic Engineering

"Human genetic engineering" refers to the genetic altering of human beings using the techniques of modern biological science. Nearly all observers have argued for caution in the development of human genetic engineering. Not all of these observers have agreed on their concerns, however.

Genotype and phenotype. Geneticists insist on a difference between the genetic makeup of an organism (the organism's *genotype*) and the characteristics that the organism displays (its *phenotype*). This distinction is based on the idea that there is no simple relationship between individual genes and their characteristics, and that genes do not *determine* characteristics. The phenotype is the result not only of the genotype but also of a complicated interaction between the genotype and the environment. Whether the interaction between the genotype and the environment *determines* phenotype is a matter of some debate. Some geneticists have argued that important unpredictable events take place during the development of an organism, so that not every change can be determined.

The goal of human genetic engineering. These points should be kept in mind when discussing the concerns regarding human genetic engineering. The general goal of human genetic

engineering is to change certain human characteristics. However, scientists are still a long way from knowing which human characteristics can be changed through genetic intervention. It may turn out that the causal relationship between some human characteristics and genes is really very complicated or is dependent upon important unpredictable events taking place during development. If this is the case, then changing the genes is not a reasonable or possible way of changing such characteristics. One of the central goals of genetic research is to understand the relationship between genotype and phenotype. It remains to be seen whether any of the characteristics that have been mentioned in the discussions about human genetic engineering—such as memory, intelligence, personality, and morality—can be changed by altering the genes. As a result, except for the treatment of certain diseases, most of the discussions so far have been vague in predicting what genetic engineering will be able to accomplish.

Ethical concerns. Human genetic engineering raises serious challenges to ethics in general and bioethics in particular. Many ethical arguments are based on differences such as society versus nature, improvements versus corrections, harms versus benefits. It is exactly these differences that may be changed by genetic engineering. Therefore, determining the morality of genetic engineering is most important. In regard to human genetic engineering, we may need to rethink many of our moral beliefs.

In a 1990 letter to the editor of *Science*, American Nobel Laureate in Medicine Salvador A. Luria (1912–1991) wrote:

[Genetic manipulation may lead to] "the possible emergence of an establishment program to invade the rights and privacy of individuals . . . to "perfect" human individuals by "correcting" their genomes in conformity, perhaps, to an ideal, white, Judeo-Christian, economically successful genotype."

GENETICS AND HUMAN SELF-UNDERSTANDING

Sherlock Holmes once asked his friend and colleague Dr. Watson, "Did you notice the remarkable thing about the dog which barked in the middle of the night?" Watson replied, "I heard no dog barking in the middle of the night." Holmes responded, "That was the remarkable thing."

And the most remarkable thing about the influence of genetics on human self-understanding is that such influence has been so limited.

Genetic Facts Ignored or Rejected

Ideas concerning heredity that in the twentieth century won many powerful believers and led to the most devastating results were developed long before there was any understanding of the gene. Nor did many believers show much concern about the fact that the developing science of genetics gave them no basis whatsoever to hold to their racial doctrines.

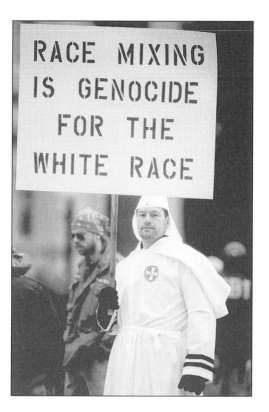

▲ A Klansman publicly affirms his belief in the genetic superiority of white people.

At the 1973 Symposium on the Identity and Dignity of Man, theologian James M. Gustafson stated:

> The most chilling thing said here this afternoon is that we couldn't afford, or ought not to permit, another potential Dostoyevsky to come into existence because one could early discern the epilepsy. . . . Are we really so sure that the qualities of normality with reference to some biological norms are the kinds of qualities that are going to make life interesting, rich and fulfilling?

Racism and heredity. Arthur de Gobineau (1816–1882) was a Frenchman who has come to be known as "the Father of Racism." His master idea was that of a supposedly superior "Aryan" people, who alone possessed true value. This people, he believed, had been contaminated by the mixing of the races and was now under threat from other white people within Europe and nonwhite people outside Europe.

These ideas gained an immediate following in Germany. British-born Houston Steward Chamberlain (1855–1927) became a German citizen just before World War I. He then became a personal friend of the emperor Wilhelm II and was recruited as an official supporter of the German cause, which Chamberlain presented as an attempt—by the nation that was "manifestly superior" to all others in art, music, literature, philosophy, and science—to take over Europe.

Chamberlain's *Foundations of the Nineteenth Century* (1899) was his best-known work. A monstrous book, it is a mixture of twisted historical "facts" and radical German mysticism. It presents the world as a stage for a conflict between an "Aryan" principle of good and the "Jewish" principle of evil. In October 1923, Adolf Hitler and Alfred Rosenberg recognized and accepted Chamberlain as the prophet of the future Third Reich. There is no reason to doubt that Alfred Rosenberg, official philosopher of the National Socialist German Workers' Party, considered his own book, *The Myth of the Twentieth Century* (1934) to have been built upon Chamberlain's *Foundations of the Nineteenth Century.*

Social inequalities versus heredity. A completely different tradition also began before genetics was seen as a science. Again the founding father was a Frenchman, although he wrote a century earlier than Gobineau. The philosopher Claude-Adrien Helvétius (1715–1771) argued in his *Concerning Mind* (1758) that "All men are born equal in mental capacity; the . . . differences that seem so conspicuous result from nothing but the inequalities in men's social condition and opportunities." The book was condemned and publicly burned. Fifteen years later, in his *Concerning Man*, Helvétius's rallying cry became "education can do everything."

The followers of Gobineau simply ignored the claims of genetic science. Helvétius's followers, however, deny completely that there are or could be any important findings about heredity in the reality of different mental capacities. As his later statement makes clear, the most important purpose of this denial was to clear the way for a program that saw the environmental factor as the only one. According to this view we are all, in all things, completely creatures of our different environments. The promise of that program was summarized in this famous quote: "Give me a dozen healthy infants, and my own world to bring them up in, and I'll guarantee to take any one at random and train him to become any type of specialist you like to select—doctor, lawyer, merchant, soldier, sailor, beggar-man or thief."

GENETICS AND THE LAW

The relationship between genetics and the law is based on the powerful information that can be known about individuals through genetic testing. In general, genetic data can be divided into two broad categories: (1) information that determines identity or biological relationship, and (2) information that can diagnose or predict one's health or the health of one's potential children. We have entered an age where genetic testing can make precise statements about identity and detect important facts about health. Therefore, it is necessary that societies develop rules to control the use of this information. The possibility that genetically engineered organisms can cause biological dangers also has encouraged much public discussion and some legal actions.

Who Should Have Access to Genetic Information?

The most important legal issue that has come out of our growing power to read the genetic code can be summarized in a single question: Who should have access to genetic information about ourselves? The answer each society comes up with will reflect the extent of the individual's right of privacy, which differs widely. In the United States, Canada, the European Union, Australia, and Japan, discussion of the need to develop rules on obtaining, storing, and controlling genetic information is under way, and guidelines are beginning to appear.

Genetic testing raises difficult privacy battles between individuals. Spouses, parents, children, siblings, and even members of one's extended family may have powerful claims to know genetic facts about an individual. For example, when a doctor determines that a child is suffering from a genetic disorder, that information may be very important to the family planning of the child's aunts and uncles. If the child's parents refuse to tell these relatives, are there situations in which the doctor should warn them? How should a doctor answer a request for genetic testing from an identical twin? The common understanding that the doctor–patient relationship will be held in confidence offers some guidance, but new questions remain. There are almost no laws concerning the responsibility of an individual to share genetic information with relatives. This topic has been looked at by several national committees that have strongly valued the protection of personal privacy, while agreeing that situations might come up that would justify sharing confidential information with a third party.

Beyond the family, genetic information may be of interest to public-health authorities, insurance companies, employers, school systems, child-welfare agencies, law-enforcement officials, the courts, and the military. In the United States, there has been discussion about whether existing legislation and common law are enough to guide the general uses of genetic data, such as in the

see also

Professional–Patient Issues:
Professional–Patient Relationship

underwriting practices of insurance companies, the hiring practices of employers, and the procedures of adoption agencies. Constitutional law, especially as it pertains to determining the limits to which a state can go to further public health, is applicable to the setting up of genetic-screening programs and the development of **DNA data banks**.

The Decision to Undergo Genetic Testing

In democratic societies, an individual has the right to decide whether or not to undergo genetic testing. The exception is newborn screening. These programs test infants for several rare, treatable genetic diseases, usually without giving parents an opportunity to refuse. In the United States, mandatory newborn screening has never been challenged in court, but would likely be upheld as a well-founded action of the public-health sector. However, state-based mandatory screening to identify children with genetic conditions that are not associated with obvious disease are less likely to become law.

Informed Consent

In optional situations, decisions about whether to undergo genetic testing are guided by informed consent, taken from laws and medical writings that have developed mostly in the United States since the 1960s.

An informed choice presumes that the person has had an opportunity to become educated about the related issues. At the very least, the doctor or genetic counselor should tell the person about the nature and purpose of the test, any significant risks associated with it, and problems that may arise from undergoing the procedure or learning the result. The individual should be told that test results could reveal that assumed biological relationships are incorrect, such as in adoption cases or when a woman is not sure who the child's father is. Some critics argue that individuals should be told that a diagnosis of disease *before* the onset of symptoms or the recognition of increased risk for a serious disorder could jeopardize their access to health and life insurance and limit their employment opportunities.

Liability for Failing to Inform About Relevant Genetic Tests

Most of the court cases concerning genetic information have focused on whether or not a doctor had a duty to warn a patient that she was at high risk for having a child with a serious genetic disorder. These court cases have led to the concepts of "wrongful birth" and "wrongful life." The concept of wrongful life came from a discussion of whether children should have the right to sue for the

DNA data bank a place where DNA information and material are stored

Genetics: DNA Testing

Ethics and Law: Information Disclosure, Truth-Telling, and Informed Consent

shame of being illegitimate. The concept of wrongful birth came from court decisions and legal articles that considered the extent of a doctor's duty to inform women about reproductive risks.

Suing for birth defects. In the United States, most state and local governments do not allow children with birth defects to bring wrongful-life lawsuits against doctors. Such suits typically accuse a doctor of failing to warn parents about a reproductive risk, which then happened. Judges have been unwilling to allow the child who is suing to argue that if the doctor had warned the parents of the risk involved, he or she would not have been born. To do so would require the jury to measure the value of an impaired life (due to a serious birth defect) against not being born at all.

In contrast, many governments do allow wrongful-birth lawsuits brought by the parents of children with birth defects. These governments believe that the doctor's failure to warn the parents violated a "duty of care" and deprived them of an opportunity to avoid the birth of an affected child. If a jury decides that such a duty exists, that the doctor violated it, and that an injury occurred, then the jury must decide the damages. Almost all governments permit the jury to estimate the special costs of raising a child with the particular disorder. Some permit additional damages to be awarded to the parents for loss of economic productivity, and a few permit damages to be awarded for emotional harm.

The availability of better genetic tests, some able to diagnose only mild to moderately severe disorders, will eventually set the limits of the doctor's duty to warn patients. It is likely that actions for wrongful birth will be limited to situations where the doctors knew or should have known of the availability of a test for parents at increased risk of having a child with a serious disease that shows itself at birth or in childhood.

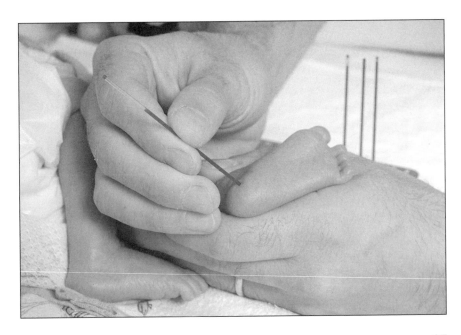

A nurse takes a blood sample from a newborn baby for a phenylketonuria (PKU) test. PKU is a genetic disorder that causes mental retardation if untreated. In the United States the test became mandatory for all newborns in 1962.

Newborn screening. In 1962, the first law to require newborn screening—for a disease that leads to brain damage and mental retardation—was enacted. Since then, there has been a steady growth of state-operated, population-based newborn screening in the Western world. In the United States, every state participates in newborn screening, most as required by state law. The programs include screening for disorders where early detection can save a child's life.

Screening laws typically require testing except for children whose parents object on religious grounds. In a few states, the parent is routinely given an opportunity to refuse the test. A report by the Institute of Medicine has concluded that newborn screening can be safely and efficiently conducted on a voluntary basis. The institute has recommended that new tests be added only when their possible benefits have been confirmed.

Carrier screening. Carrier screening is meant to identify otherwise healthy individuals who, depending upon whom they marry, may be at high risk for having children with a severe genetic disorder. (Such individuals are called carriers because, while they don't have a genetic disorder themselves, they "carry" the gene for one.) In the United States, population-based screening programs have been set up to identify carriers for sickle-cell anemia and Tay-Sachs disease. Since its start in the mid-1970s, Tay-Sachs screening has been community-based, performed without state interference, and used often by the Eastern-European Jewish population in which the disease is relatively common. On the other hand, sickle-cell screening programs were often started by the government. Between 1970 and 1972, twelve states and the District of Columbia enacted laws that required carrier screening of African Americans, usually by tying it to entry into public schools or to obtaining a marriage license. Most state laws failed to offer pretest education or access to genetic counseling. Concerns about genetic discrimination provoked a strong outcry from the African-American community and led to new federal laws.

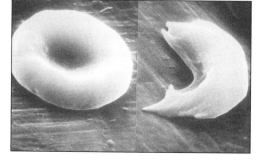

A normal and a sickle-cell red blood cell, enlarged 18,000 times.

Laws to End Genetic Discrimination

Concerns about genetic discrimination as a result of screening and testing have grown among geneticists since the mid-1980s. The possibility that genetic data could be used to deny individuals access to health insurance has become a major topic.

In 1989, the Human Genome Privacy bill was introduced in Congress. It was reintroduced in 1990, and hearings were held in 1991. But the bill, which was concerned only with the protection of genetic data obtained through the use of federal money, did not become law. Since 1991, several states have considered, and some have adopted, laws that specifically forbid the use of genetic data in certain situations, usually involving life insurance or job hiring.

Wisconsin forbids employers from requiring or performing a genetic test without first getting the subject's written informed consent. California forbids insurance companies from charging higher rates to individuals because they carry a disease-bearing gene that may be associated with disability in their future children. Montana forbids life-insurance companies from refusing applicants on the basis of a "specific chromosomal or single-gene genetic condition."

The Americans with Disabilities Act (ADA) of 1990 covers individuals who can show that they have been discriminated against because they are seen as being disabled. This law has raised the possibility that otherwise healthy individuals who carry a gene that will eventually cause a disease or increase their chances of becoming ill may seek compensation due to employment discrimination. The courts will, in the end, decide the range of coverage. One court has held that severe obesity, a genetically influenced condition, may qualify for protection under the Rehabilitation Act, which uses the same definition of disability as does the ADA.

Genetic discrimination in determining access to health insurance is a major problem only in the United States and in South Africa—two nations that have not provided their citizens with basic health care.

see also
General Topics: Disability

GENETIC TESTING AND SCREENING

embryo a developing unborn baby (from fertilization to about eight weeks)

see also
Fertility and Reproduction: Fetus

This article focuses on two types of genetics testing and screening: preimplantation diagnosis and prenatal diagnosis.

Preimplantation Diagnosis

Preimplantation diagnosis—the detection of genetic defects that cause inherited disease in human **embryos** before implantation—has many benefits for couples known to be at risk of having affected children. The main benefit is that the ability to select unaffected embryos for transfer to the uterus (womb) means that any resulting pregnancy should be normal. It would make it no longer necessary to terminate in the later stages of development a pregnancy diagnosed as affected. Another benefit relates to the use of superovulation and in vitro fertilization (IVF). Superovulation increases the number of eggs that develop in a single reproductive cycle. After IVF, an average of five or six embryos can be screened at the same time. This increases the chance of identifying unaffected embryos. Although several IVF cycles may be necessary, establishing a normal pregnancy may in many cases take less time than terminating affected pregnancies diagnosed at later stages.

For DNA analysis, cells are removed from each embryo at about the eight-cell stage, early on the third day after insemination. The DNA analysis is then carried out, if possible within eight to

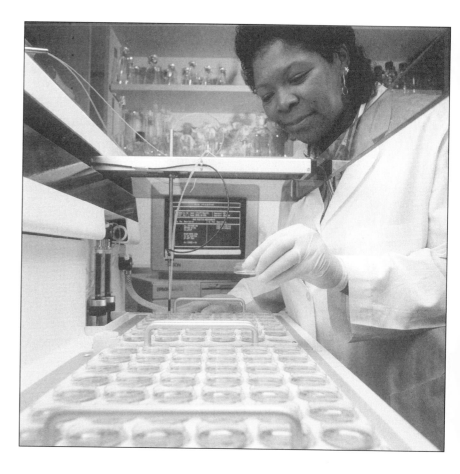

▶ Rows of petri dishes contain cells collected through amniocentesis to determine characteristics of a human embryo.

twelve hours, and unaffected embryos are transferred to the uterus later the same day. The embryos are apparently unharmed by this process.

Pregnancies (and in some cases births) have been achieved mainly after preimplantation diagnosis to identify the sex of embryos and, in the case of couples at risk of diseases affecting only boys, to transfer females. However, specific diagnosis of embryos affected by cystic fibrosis and several other diseases has also been achieved.

Prenatal Diagnosis

Modern prenatal (before birth) diagnosis began in the late 1960s and early 1970s with the development of laboratory techniques that allow **amniotic fluid** cells to be grown in a laboratory.

amniotic fluid the fluid in the uterus surrounding the embryo or fetus

From 2 to 4 percent of all infants are born with a serious defect. If minor defects are included, the rate can reach 8 to 10 percent. Approximately half of all serious defects are genetic. The other half can be connected to drugs, alcohol, infections, or other nongenetic causes.

The following are reasons for offering prenatal diagnosis: (1) advanced maternal age—at age twenty-five, a woman has a 1 in 476 risk of having a baby with a chromosome abnormality; at age thirty, the risk is 1 in 385; at age thirty-five it is 1 in 192, and at age forty it

is 1 in 66; (2) a genetic history of abnormalities; (3) belonging to an ethnic group known to be at risk; (4) previous poor reproductive outcomes, including infants born with birth defects; (5) a family history of infants with birth defects; (6) more than one miscarriage; and (7) a possible problem in a current pregnancy.

Advanced maternal age is still the most common reason for prenatal diagnosis. Worldwide, it accounts for more than 90 percent of all prenatal tests. Although it is common care in the United States to inform pregnant women over the age of thirty-five that prenatal diagnosis is available, the choice of that age as the cutoff is subject to judgment. In fact, the risk of having a child with a chromosome abnormality begins to increase before the age of thirty, then continues to rise at a fast rate.

Couples who already have had a child with Down's syndrome or another chromosomal abnormality face a higher risk in each following pregnancy. They are candidates for prenatal diagnosis regardless of age.

A known biological abnormality in one of the parents is a less frequent but important reason for prenatal diagnosis. Some abnormalities come to light only after the birth of a child with an "unbalanced" chromosomal rearrangement or as part of the diagnostic evaluation after multiple early miscarriages. Other abnormalities are linked to certain ethnic groups.

Procedures for Prenatal Diagnosis

There are three basic approaches to identifying fetal defects: visualization, analysis of fetal tissues, and laboratory studies (see p. 33).

Visualization. The use of ultrasound in the study of human anatomy has had a huge impact on all areas of medicine. The

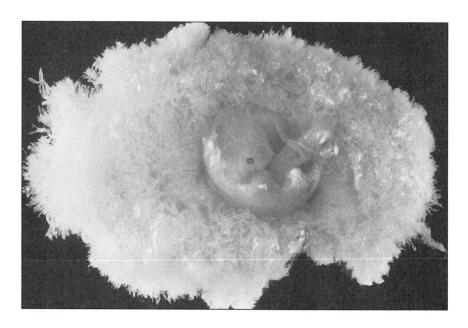

A tiny six-week-old embryo nestled within the chorion (embryonic membrane). The projections on the surface of the chorion (villi) can be examined in a procedure aiming at early detection of genetic defects.

development of high-resolution, real-time scanners has made it possible to see almost any part of the fetus in great detail.

Ultrasound exams are performed in 30 to 50 percent of all pregnancies in the United States, often in the doctor's office. More detailed examinations are best done by professionals specially trained in ultrasound exams. Prenatal diagnosis may be most helpful when the doctor looks for a particular problem. A thorough examination includes many different measurements to determine the growth and the age of the fetus. In addition, each part of the fetus's anatomy—face, chest, heart, abdomen—is examined in detail.

Analysis of fetal tissues. Visualization techniques, such as ultrasound examinations, do not provide genetic information about the fetus. In order to obtain this type of information, amniocentesis or chorionic villus sampling may be required when a fetus is considered to be at risk for genetic abnormality.

Amniocentesis. Puncturing the uterus with a needle to draw out some of the amniotic fluid is the most often used technique for prenatal diagnosis. This type of amniocentesis can provide the doctor with a wealth of information about the condition of the fetus.

Amniocentesis should be done by a health professional trained in the technique. It is often performed with the help of ultrasound. Usually a sufficient sample of amniotic fluid is obtained on the first try, and successful cultures are created in almost all cases. The greatest risk of this test is the possibility of miscarriage. In the 1976 National Institute of Child Health and Human Development study, the loss rate was approximately 1 in 200. By the late 1990s, that rate was believed to have decreased in major centers to about 1 in 300.

With more experience and confidence in performing the test, it has become possible to perform amniocentesis earlier and earlier. Several prenatal diagnostic programs have begun offering amniocentesis as early as ten to fourteen weeks into the pregnancy. The risks of performing this test early in the pregnancy are not yet fully known.

Chorionic villus sampling. Although amniocentesis has proved to be very safe and highly reliable, a test that can provide information before eighteen to twenty weeks into the pregnancy is still preferred. Although the decision to end a pregnancy is rarely an easy one, it is perhaps more easily made earlier in the pregnancy.

An alternative test to amniocentesis, chorionic villus sampling can be performed during the first three months of pregnancy. The chorionic villi are the forerunners of the **placenta** and can be drawn out of the uterus, usually between nine and twelve weeks into the pregnancy.

As with amniocentesis, there has been great concern about losing the fetus or seriously damaging it. It is difficult, however, to determine exactly the rate of risk because it is difficult to separate the possible causes—the test itself, the age of the mother, and other factors such as tobacco and alcohol consumption.

Amniocentesis was first tried in 1882 and has long been used in late pregnancy to find out if the fetal lungs are mature enough for delivery. In the latter part of the 1970s, amniocentesis came into use in the second trimester of pregnancy to diagnose certain birth defects. Ultrasound is used to determine the exact locations of both the fetus and the placenta, thereby allowing the technician to choose the safest spot for inserting the needle. The cells in the amniotic fluid sample are then grown in the laboratory for one to two weeks and tested for chromosomal abnormalities or various genetic birth defects. Test results are usually available within two to three weeks.

placenta the organ in most mammals that joins the fetus to the mother's uterus; through the placenta the fetus is nourished and its wastes are removed

---

Methods Used for Prenatal Diagnosis

I.   *Visualization*
Noninvasive
    Ultrasonography
Invasive
    Embryoscopy
    Fetoscopy
    Endoscopy

II.   *Analysis of Fetal Tissues*
Amniocentesis
Chorionic villus sampling
Cordocentesis (fetal blood sampling)
Skin biopsy
Liver biopsy
Muscle biopsy

III. *Laboratory Studies*
Cytogenics
Biochemical
DNA

 Canadian scientist Oswald Avery was first to show that genetic information is transferred by deoxyribonucleic acid (DNA).

Ethical Connections

It has become a basic belief of genetics that there is no link between offering prenatal diagnosis, documenting fetal abnormalities, and deciding to have an abortion. *Prenatal diagnosis is for the purpose of providing information to couples about what they can expect.* Even for couples who would never consider abortion as an option, the knowledge obtained from prenatal diagnosis can be very valuable. The level of anxiety can be greatly lowered when a normal result is found. On the occasion that an abnormality is found, the parents then have the chance to prepare for what is coming, and maybe change the course of prenatal care.

The use of screening tests, such as ultrasound, has brought many couples to prenatal diagnosis who otherwise would never have been seen. This is important to the general public health because more cases of genetic problems can be found. Advanced tests were once regarded by many women as being unnecessary and scary, but they have come to be more readily accepted. In most cases, these patients' anxiety is reduced when they hear that their pregnancy is normal.

Obstetricians (childbirth doctors) have long believed that there are two patients—the mother and the fetus. Only with the discovery of ultrasound and fetal treatments has the distinction

between mother and fetus become generally understood—raising the status of the fetus ethically and sometimes legally. Understandably, there are sometimes conflicts of interest between the two that have serious legal and ethical implications.

GENOME MAPPING AND SEQUENCING

chromosome a rod-shaped body that contains genes of an organism; different organisms have differing numbers of chromosomes to hold all the genes

gene one unit in a chromosome that determines a trait

genome all the genetic material of an organism; a complete set of chromosomes of an organism with the genes they contain

Genetics: DNA Typing

mutation a significant and basic change in living organisms

natural selection the theory that animals and plants best adapted to their environment tend to survive and that future generations show an increasing adaptation to that particular environment

Discoveries in the field of human genetics have had an enormous impact on biomedical research. At the center of these discoveries was the development of new techniques for making maps of human **chromosomes**. Before the 1980s, when most of these techniques were developed, scientists could only look for one **gene** at a time. For instance, they would try to find the gene that caused a certain disease. But with the invention of genetic mapping, scientists could look at all the genes on a set of chromosomes. A genetic map is like a street map, but instead of presenting a picture of the streets in a particular town, it provides information about a chromosome or set of chromosomes. It can determine where on a chromosome a particular gene lies. The new mapping techniques allow scientists to map an entire **genome**. This means they can know the location of all the genes on each chromosome. The new techniques that allow scientists to make a map of the genome also allow them to study the material that genes and chromosomes are made of—DNA.

Historical overview. Genetics was not a central field of scientific study until the twentieth century. In fact, the term itself was not used until around 1900. The study of genetics grew out of the observation that organisms—people, animals, plants—pass along certain characteristics, or traits, from generation to generation.

During the first part of the twentieth century, many scientists found that genes are located in chromosomes. Chromosomes can be found in the nucleus of every cell in the body. They are very small, but they can be seen with the help of a microscope. Scientists found that they could look at chromosomes by placing a group of cells on a slide and staining them with a special dye. When chromosomes were discovered, no one knew what chemicals they were made of. No one knew what the physical makeup of a gene was. The term "genome" refers to all the genetic material in the chromosomes of each species of animal or plant. When reproduction takes place, genes are sorted differently among offspring. This means that not all offspring will get exactly the same genes from their parent. However, once the genes are inherited, they usually do not change. The only way that a gene will change is by **mutation**. This means that for the most part, from one generation to the next, a species will have the same characteristics. Every once in a while a mutation does take place, and this might cause the species to change. The mutation could make the species better able to live in its environment. Gene mutation and gene "shuffling" are how **natural selection** can take place.

A model of the DNA double helix structure discovered by James Watson and Francis Crick.

In the middle of the twentieth century—the 1940s and 1950s—scientists moved beyond understanding how genes allow traits to be inherited. They began to understand the physical makeup of genes. In 1944, the scientists Oswald Avery, Colin MacLeod, and Maclyn McCarty discovered that *deoxyribonucleic acid* (DNA) is the chemical that genes and chromosomes are made of. Through their experiments, they found that DNA is what confers inheritance. About ten years later, in 1953, two other scientists, James Watson and Francis Crick, discovered the physical structure of DNA. They found that DNA is made up of a long chain of four constituent bases. The names of these bases are adenine (A), cytosine (C), guanine (G), and thymine (T). Millions of these bases are linked together in long chains that are twisted and folded into chromosomes. Along these chains, the bases are combined in infinitely different ways, and this is how DNA makes up the different genes that cause different traits. The bases are linked together by a sugar molecule and a phosphate molecule. The sugar and phosphate molecules make up the "backbone" of the DNA chain.

The Hierarchy of DNA

1. Nucleotides: Adenine (A), guanine (G), cytosine (C), and thymine (T), the basic building blocks of DNA.

2. Base Pairs: Nucleotides are found paired together (A–T, T–A, C–G, G–C) in the shape of a helix (a twisted ladder).

3. Codons: Nucleotides arranged in a triplet code, each corresponding to an amino acid (components of proteins) or a regulatory signal.

4. Genes: Functional units of DNA needed to synthesize proteins or regulate cell function.

5. Chromosome: Thousands of genes arranged in a linear sequence, consisting of a complex of DNA and proteins.

6. Genome: The complete set of genetic information on all 46 human chromosomes.

Gene Maps

For many years, scientists who studied genetics were mostly interested in finding out which genes cause certain diseases. The process was very long and involved. The scientist had to go through a series of steps. Each step required a different kind of genetic map. In general, all genetic maps show the sequence of DNA bases that make up a particular chromosome and tell where on the chromosome a gene is located. But there are several kinds of maps and each one is used for a different purpose.

▶ James Watson and Francis Crick, discoverers of the DNA structure, with a tall model of the DNA double helix (1953).

genetic marker a gene or trait that serves to identify other genes or traits linked with it

Genetic linkage maps allow a scientist to look at the relationship between different genes. A linkage map can show if two different traits that are located in a chromosome will always be inherited together. Linkage maps use **genetic markers**, which contain known DNA sequences. For instance, a genetic marker may contain the following standard sequence:

A–T–A–T–G–G–C–C–A–A–T–A.

Scientists can look for this and other sequences on a chromosome. A linkage map will show the scientist where in a chromosome this sequence appears, where variations in the sequence appear, and the relationship between these sequences on a chromosome. A map of these variations can be used to follow the inheritance of parts or regions of chromosomes. A linkage map could show that the sequence above is located on the left leg of chromosome 8. The chromosome region can then be matched up with the inheritance of a particular characteristic or disease. In this way, the rough location of the gene that causes the trait or disease can be discovered.

Genetic linkage maps have helped scientists make important discoveries about inherited diseases. If the children from different families inherit a certain disease along with a certain region of a chromosome, the disease is probably caused by a gene on that part of the chromosome. One early genetic mapping experiment was able to locate the gene that causes the disease cystic fibrosis. These experiments found that the gene is located on chromosome 7. Genetic mapping showed that children with cystic fibrosis inherited one version of chromosome 7 from their parents while their healthy brothers and sisters inherited another.

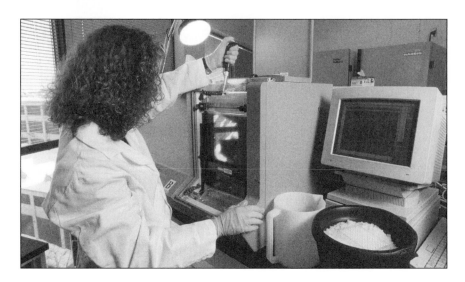

A technician loads a gene mapping machine.

Genetics is both the study of the characteristics that run in families and the chemicals and molecules that make up genes. Genetic linkage maps show that human genetics and molecular genetics are related.

In 1980, American scientists published an important paper in the study of genetics. The authors explained that the kinds of genetic differences between individuals that can be seen in a linkage map could be used to make a map of the entire human genome. A map could be assembled by conducting experiments that would search for minor differences in DNA sequences. The amount of variation would be very small compared with the amount of similarity but, according to the authors, even these tiny differences could be observed.

Physical mapping is used to study the DNA from a certain region of a chromosome. The most useful form of a physical map is a set of ordered clones that contain DNA spanning an entire region of the chromosome. The chromosomal DNA is broken into small fragments that can be copied in bacteria or yeast (the genetically identical clones). The fragments are copied many times and then put back together.

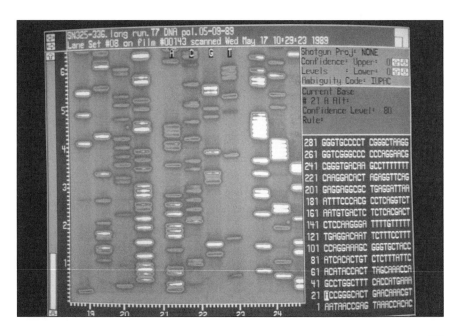

DNA sequencing gels depicted on a computer screen, 1989.

DNA sequencing, a type of physical mapping, is the most detailed map form. DNA sequence is, in a genetic sense, the territory to be mapped. The exact order of the chains of A-C-G-T bases is what provides the most information about chromosomes. In the mid-1970s, two different groups of scientists discovered how to determine the sequence of DNA bases. As procedures became faster and less expensive, DNA-sequencing techniques gave scientists a very powerful tool. They could find out everything they wished to know about the genome. Then, three individual scientists proposed mapping the DNA sequences of the entire human genome. (Each scientist wanted to get a model reference sequence that could be used for study, not the entire genome of a particular individual.) Proposals to sequence the human genome sparked a controversy.

The Human Genome Project

The debate that followed led to the creation of the Human Genome Project. The original proposals for DNA sequencing were broadened to include physical and genetic linkage mapping; studying the genes of nonhuman organisms; and developing new techniques for analyzing DNA and for interpreting the results of gene-mapping experiments. The Human Genome Project produced maps of the human genome as well as the genomes of other animals. It also developed new ways of getting information about DNA sequences, focusing on automation and robotics.

Genetic analysis goes well beyond finding out about the diseases inherited according to simple single-gene patterns. Molecular genetics can provide information about diseases that no other techniques can. Without these molecular genetic techniques, scientists would know very little about some very common diseases passed along by genes. Some examples of genetic conditions include hardening of the arteries, arthritis, Alzheimer's disease, high blood pressure, diabetes, and many kinds of cancer. All these diseases can run in families. When a hereditary disease can be explained by the presence of a single gene, finding that single gene is the first step in understanding how the disease works and how it might be cured. A complete genetic map could help scientists identify the genes that cause most single-gene inherited traits. This kind of research requires many repetitive experiments. Complete genetic maps also would allow scientists to search for two or more genes that together cause a disease. High blood pressure and diabetes are two diseases that appear to be caused by many genes working together.

Social Policy Issues

Government leaders and policymakers welcomed the new techniques and discoveries in genetics. They hoped that these new

▲ Dr. Margery Shaw and Dr. T. C. Hsu of the University of Texas developed a process of staining certain genes within chromosomes, making it possible to discover potential birth defects and hereditary illnesses (1971).

Genetic disorders are transmitted from parents to their children in several ways:

• In *dominant inheritance*, one parent has a single defective gene that dominates the corresponding normal gene in the pair. Common disorders in this category include glaucoma, which can cause blindness.

• In *recessive inheritance*, both parents carry the defective gene (although they may not be aware of it). Recessive disorders are rare in the general population, but some of them may affect specific ethnic groups in which a particular gene may be unusually common. Tay-Sachs disease is found in Jews of Eastern European descent, and sickle cell anemia occurs more frequently among African Americans.

• In *sex-linked inheritance*, a defective gene on one of a woman's X chromosomes may cause her male children to get the disease carried by the faulty gene. A daughter will not get the disorder but, as is the case with hemophilia, she may inherit the faulty gene and transmit it to her male children.

discoveries would increase understanding of many diseases. The public also strongly supported biomedical research. Certainly, being able to map the human genome can make biomedical research better and easier. But along with the strong support, many people responded to the big changes in genetic research cautiously.

No one was really worried about the way the research itself was being conducted. But both policymakers and the public expressed concern about how the new techniques for understanding the composition of the genome and the knowledge gained might be used. The most obvious effect of techniques for mapping genes and sequencing DNA was to speed up the advances in the field of human genetics.

Threat to Privacy

One of the concerns was that these new technologies, which could tell us a great deal more than ever before about an individual's genetic makeup, might be used for testing and screening purposes. The origin of this concern lay in the 1960s and 1970s, when tests were developed that could screen people for two genetic diseases, phenylketonuria and sickle-cell anemia. It seemed as if genetic screening might pose a danger to people's right to privacy and to fair employment. People worried that if their doctor found out that they had a gene for a particular disease, this information might not be kept private and might be used against them. If genetic screening revealed that they carried a certain kind of gene, it might be more difficult for them to get a job, health insurance, life insurance, and other social benefits. A health insurance company might see them as too much of a risk. An employer might not hire them because they could fall ill in the future.

Modern molecular genetics is very different from eugenics. Eugenics concentrated on family background, bone structure, and other superficial factors. Molecular genetics is a much more scientific approach to understanding and predicting the results of human differences than eugenics. However, the public is right to be concerned. It is important to question the interpretations of scientists, especially regarding human genetics. The information that modern genetics supplies can be just as easily misused to control and limit individual freedom. If gene-mapping techniques can provide very accurate information about important aspects of persons, such as their ability to play sports or their intelligence, there is a danger that the information can be used to categorize and exclude people based on their genetic makeup.

ALLOCATION OF HEALTH-CARE RESOURCES

"What is the use of discussing a man's abstract right to food or medicine? The question is upon the method of procuring and administering them."

—Edmund Burke (1729–1797),
British statesman

In any health-care system, there is a limited amount of money, services, and other resources, and someone must decide how they are divided up. Allocation refers to decisions about how health care is distributed or provided to various people. It would seem most efficient to have a particular individual or group whose job it is to make these decisions. In the United States, though, there is no such system. Who gets health care, what services they get, how health-care money is spent, and how resources are divided are decided by millions of different health-care providers and patients, and by conditions in the health-care market. If you are sick, there are no official guidelines about what treatment you must receive. You and your doctor will discuss the problem and decide what treatments are best, based on what you need and can afford. Your treatment might also be affected by whether or not you have health insurance and what kind of care the insurance pays for.

Microallocation

Health-care allocations can be divided into two different levels of decision making: microallocation and macroallocation. Microallocation focuses on how scarce health-care resources are divided up among particular individuals. If several patients need a heart transplant but there is only one heart available, how do doctors decide which patient gets the transplant? Microallocation can also involve making decisions about one person's care. If a person has a disease and there are several different treatments available, how do doctors decide which treatment to use?

Macroallocation

Macroallocation involves decisions that are made at a higher level and that affect larger groups of people. These decisions determine the amount of resources that are available for different health-care services. Macroallocation might involve a hospital's decisions about what kinds of care it will provide and where it will spend its money. Or it might involve a nation's decision about how much money is spent on high-technology medicine and how much is spent on preventing childhood diseases. Another example of macroallocation is a nation's decision to spend time and money on cleaning up the environment, since that also affects people's health. A nation might also decide to spend most of its money on education or defense instead of health care.

▶ Shortly after her husband was elected president of the United States, Hillary Rodham Clinton began her campaign for a national system of health care that would cover all Americans.

Bioethicist Daniel Callahan wrote a book entitled *Setting Limits: Medical Goals in an Aging Society* (1987) in which he argued that life-extending medical treatments should not be provided for individuals older than eighty. The health-care system should instead concentrate on providing them with good long-term care and nursing-home support. Callahan's book raised a storm of controversy. However, he was extremely brave to begin a serious American debate about what forms of health care we can afford to provide for the very old, especially when younger generations have important competing needs.

Cross-subsidization: Galen (130–220 C.E.) was a famous Greek doctor. He believed that doctors should heal the sick primarily because of love of humanity (*philanthropia*), and not because of the love of money. Galen accepted payment only from his wealthy patients.

Rationing

Rationing involves deciding how health care is divided up when there are not enough resources for all patients to get the necessary care. When health care is rationed, some people will be left out and will not get the care they need.

Some use the term "rationing" only when there is a clear decision to leave someone out of care. These critics are often against rationing because they believe there are, or should be, enough resources for everyone to get basic care. Others use the term when care is not available for any reason, even if no one has made a clear decision not to provide care. Many in this group believe that resources will always be limited and that this will always result in some people not getting the care they need. They believe that we should develop clear ethical guidelines for deciding who gets care when there is not enough for everyone.

Because different people use the word differently, "rationing" is not a very clear term. It is also misleading because many people think of rationing as a temporary policy. For example, during war, things like gasoline, meat, or sugar are sometimes rationed, so that each person gets a fair share. After the war, these things are not rationed any more, and people can buy as much as they want or can afford. In health care, rationing is not a temporary situation but a long-term decision.

The debate over rationing and what it means in health care has brought up two important questions: (1) Are some people being left out of health care? (2) In the future, will we need to make clear decisions about who does or does not get care?

It is clear that some people are being left out, even if it is not done on purpose. In Canada and Europe, people often must spend time on a waiting list before they can receive certain kinds of care. In these countries, no one wants to make people wait or keep them out of care, but this happens anyway because of government limits

William Osler (1849–1919) practiced medicine in Philadelphia and Baltimore. He did not charge poor patients, but required unusually high fees from wealthy patients. Until recently, many hospitals provided free service to the uninsured by charging higher prices to the wealthy and to insurers. But this practice has been disappearing.

on health-care spending, the structure of the system, and other features. In underdeveloped countries, most people are left out of health care because the hospitals are located in the cities, and people who live in the country may be too far away and too poor to travel to the city to get care.

Related Literature

Chaucer noted in the *Canterbury Tales* that physicians have a "special love for gold."

ECONOMIC CONCEPTS IN HEALTH CARE

economic related to the distribution of wealth

economics the study of how people produce, distribute, and use money, labor, and resources

commodity something that is bought and sold

Health care has always been an **economic** activity. People invest time and other resources in it, and they trade for it with one another. In this way, health care is open to economic analysis. This means understanding the demand for it, its supply, its price, and the relationship between all three. Economic analysis does not merely determine what the supply, demand, and price for health care in private or public markets are. It also tries to understand why those are what they are. How does suppliers' behavior affect the demand for health care? How does a particular insurance framework affect the supply and demand?

The economics of health care, in fact, has grown into a specialty within professional **economics**. Although almost all goods are in some sense "economic goods," economists have noticed some differences between most health-care and ordinary market **commodities**. In health care, the supplier seems to create a final demand more than with most goods, since both the shape of health services and their price are directly influenced by health-care providers. Other forms of "market failure" also occur in health care. An example of this is when people do not receive health-care services because their high risk to insurance companies drives prices for even the most basic coverage to levels that no one can afford.

Cost-Effectiveness, Cost–Benefit, and Risk–Benefit Analysis

Efficiency involves the basic economic idea of "opportunity cost." This is the value from alternatives that we might have chosen using the same resources. When the value of any alternative is less than the value of the current service, the current one is efficient. When the value of some alternative is greater, the current service is inefficient. If we focus on whether we can get more benefit for our health-care dollars, or whether we can get the same health care more

cheaply, we are performing cost-effectiveness analysis (CEA). But if we compare an investment in health care with all the other things we might have done with the same time, effort, and money, we are performing cost–benefit analysis (CBA). CEA asks whether the money spent on a particular program or treatment could produce healthier or longer lives if it were spent on other forms of care. CBA asks an even more difficult question: whether the money we spend on a particular portion of health care is "matched" by the benefit. In other words, could the money produce greater value, not just healthier or longer lives, if it were spent on other things?

Cost–effectiveness analysis. CEA is the less difficult activity: We compare different health-care services and find either final differences in expense to achieve the same health benefit or differences in some health benefit. But in CBA, we must compare the value of the longer life and better health achieved by health care with the value of the other possible improvements in human life that would be gained by making other investments with the same resources.

Cost–benefit analysis. CBA is difficult because the advantages gained from those other uses of resources so often cannot be compared to benefits in health and in living a longer life. But improvements within health care often cannot be measured either. How do we really compare the value of hip replacement, which will not make a person live longer, with that of kidney dialysis or organ transplants, which will?

Risk–benefit analysis. This is another kind of analysis: We compare the probabilities of harm presented by a certain course of action with its likely benefits. If another procedure is likely to produce almost the same benefits with less risk, then that is obviously better. But it is not always clear when one risk is "less" than another. One risk may be of paralysis for life and the other may be of ongoing pain. Moreover, one procedure may have lower risk but also promise fewer health benefits. Again we are left with a final trade-off that cannot be measured. Unlike CEA, the positive effects in risk–benefit analysis are not all measured on the same level. And unlike CBA, the benefits are not put in the same terms as the costs or risks. The analysis helps us only to see what risks we take on in the pursuit of what benefits.

Other standards. Unlike CBA, CEA does not try to answer the question of how much money to spend for a given health benefit. But it does attempt to make comparisons within health care. All that it needs to be able to do this is to determine a common unit of health benefit. One idea developed for this purpose goes by various names: a "well-year," a "quality-adjusted life year" (QALY, pronounced to rhyme with "holly"), or "health-state utility." In any case, the basic idea is a common unit that combines longevity with quality of life considerations—such as a year of healthy life. We can then compare not only actions that extend life with each other, but also actions that improve the quality of life with those that prolong it. A

Transplants and Other Technical Devices:
Kidney Dialysis

"The cost–benefit analysis approach to whether we live or die is generally unappealing."

—John Sullivan, editor, *National Review* (1989)

General Topics: Quality of Life

hip replacement, which improves the quality of life, can be compared with kidney dialysis, which prolongs it. In public-health terms, we could also track the health of a population, calculating increases (or decreases) in years of healthy life.

The major moral argument for using both quality of life and lengthening life as standards for getting the most benefit that a plan or an entire health-care system produces is that it is people themselves who rank the quality of their lives. It is also people who agree to the priorities that QALYs or well-years bring. Critics believe, however, that increasing years of healthy life in our lifesaving policies fails to respect the person with an admittedly lower quality of life.

The Monetary Value of Life

How much is your life worth? In contrast to CEA, CBA asks us to equate an amount of money with the benefits of the program or procedure that is being assessed—including life itself. Putting a price on life is an ordinary event of modern times. Now that many effective but often costly means of preserving life are available, we often pass up potential lifesaving measures for other choices. And money eases those trade-offs.

Discounted future earnings. Economists have developed two main models for translating data into an economic value of life. These models are discounted future earnings (DFE) and willingness to pay (WTP). DFE looks at the future earnings that are lost when a person dies. Economists feel it would be self-defeating to refuse to save a life for $200,000 if the value of the person's earnings was more than that. While such DFE calculations are still used in some health-care CBAs, the idea of discounted future earnings has been largely surpassed in the work of economists by WTP. With WTP, the value of life is measured by people's willingness to use resources for increasing their chances of survival.

Willingness to pay. In economics, WTP is thought of as being the better model. It captures the range of life's individual, elusive values that DFE ignores. People often spend money on something they like independent of the return they will get from it. This is the case with WTP. But WTP has raised many objections nonetheless. For one thing, just as with DFE, there are wide variations in willingness to pay, largely based on people's wealth and income. Should we see those variations as affecting what is spent on saving a life?

The move that economists make in WTP to get from an initial trade-off between money and risk to the value of a real, unique life is puzzling. John Broome (1982) claimed that only a money value made directly in the face of death can correctly reflect the actual economic value of a life. But as Ezra J. Mishan (1985) noted, we do not know of anyone "who would honestly agree to accept any sum of money to enter a gamble in which, if at the first toss of a coin it came down heads, he would be summarily executed." Some economists

> "Whatever the costs are, we can't measure health and lives in economic terms."
>
> —Thomas Luken,
> U.S. Congressman, 1990.

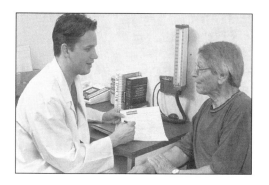

A medical assistant explains the charges on a medical bill.

conclude from this that CBA can set no reasonable limit on what to spend to save a life because no amount of money adequately represents the real value of life.

The money value placed on life brings up the idea of insurance. In medical economies, most people either subscribe to private insurance plans or are covered by public health-care spending. Once insured, subscribers and patients, as well as health-care providers, find themselves with a strong incentive to overuse health care and a tendency to underestimate opportunity costs. Why should we not address the problem of controlling the use of health care at the point in the decision process—insuring—where the cost-expansion trouble starts?

The Difficulties That Economic Concepts Pose for Clinical Practice

Suppose that economic efficiency analysis—such as CEA or CBA—lays the groundwork for recommendations about the kind and amount of health care to use. How does such an efficiency system relate to the ethical duties of those people who provide health care? The traditional code of doctors is one of loyalty to their individual patients. In turn, that means doing whatever helps a patient most, within the limits of what the patient willingly accepts. If health care is to be rationed to control the money and other resources it uses, will the basic ethical code between doctor and patient have to change? If it will, is the achievement of efficiency worth its moral cost? This potential clash between traditional ethical obligations and the economic and social demands of the "new medicine" in an age of scarce resources will be the focus of ethical controversies in medicine for years to come.

One can divide the potential views surrounding the controversy into two camps: those who think that the economic-efficiency demands of society cannot be reconciled with the ethical obligations of doctors, and those who think that they can be. The first group will go after economic efficiency regardless of the moral cost. The second group will oppose rationing of health-care services in the name of a moral commitment to helping individual patients. The views of this second group will come in distinctly different varieties: (1) the view that the controversy was more apparent than real all along, since providers of health care have always shaped the lengths to which they would go to help individual patients; (2) the position that distant third parties make all the rationing decisions and that the doctors then ration to patients within determined guidelines; and (3) the view that a provider's loyalty to a patient, though not controlled by efficiency, is that of a member of a just society, a condition that then allows the doctor to participate in the rationing of health-care resources with a clean conscience when it is based on ideas of fairness and justice.

Health Care: Allocation of Health-Care Resources.

Professional–Patient Issues: Medical Codes and Oaths, Professional Ethics, Professional–Patient Relationship

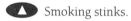
Smoking stinks.

secondhand smoke tobacco smoke that is
exhaled by smokers and inhaled by people
nearby

General Topics: Substance Abuse

Externalities and Public Goods

Externalities and public goods play an important role in discussions of public policy. Externalities are burdens, costs, or benefits that are added onto other people, not to the people who are performing some action. Externalities pose a problem for achieving efficiency in market exchanges.

Public goods raise questions of public regulation and taxation. To an economist, a "public good" is one whose benefits extend even to those who do not buy it. If you clean up your yard, I benefit from a somewhat better appearance on the block regardless of whether I clean up my own yard or help you clean up yours. The benefit is thus public. It is difficult if not impossible to exclude from the benefits of such activities someone who chooses not to contribute. The obvious solution to this unfairness on my part is for the community to tax me my fair share.

The use of both public goods and externalities is on the rise in certain health-care issues. One is the taxing of health-complicating products such as tobacco and alcohol. Smoking and too much drinking increase certain costs to others. These costs include health-care expenses for smoking- and drinking-related diseases; lost work time; unhappiness; and pain in dealing with other people's destruction of their lives. Even direct loss of life (from **secondhand smoke** and drunk driving) is often passed on to other people. These external factors provide part of the force behind the movement to raise taxes on tobacco and alcohol. As societies look for ways to deal with rising health-care costs, a fair source of revenue (income) would be from special taxes on activities that increase the health-insurance premiums and taxes of others through voluntary behavior.

In the case of smoking, the picture is very complicated. Informal cost analysis tells us that smokers cost nonsmokers a great deal of money. But that conclusion ignores two hidden "savings" of smoking that extend to others. Because smokers die earlier, and usually at the end of their earning years or shortly after retirement, they save others the pension payouts and unrelated health-care expenses they would have incurred had they lived longer. One leading study found that the average 1989-level U.S. cigarette tax of $.37 per pack was enough to cover all the costs that smokers impose on other people. Of course, this is not the last word on the external costs of smoking. But it illustrates the hidden costs and savings that economic analysis reveals.

Many economic challenges to health care will continue to surround us. Questions such as how to determine what is efficient in the investment of resources in health care, how to arrange efficient use of care, and how the achievement of efficiency compares with other health-care values will become the economic focus of the whole health-care system.

Related Literature

James Wright's "The Minneapolis Poem" (1968) records the experiences of the nameless poor who live on the streets and do not have the money to get enough to eat, let alone afford any kind of health care. Poor old men die of hunger and exposure in the cold winter nights. A companion poem, "In Terror of Hospital Bills" (1968), is told in the voice of a homeless Sioux Indian, hungry and desperate on the winter streets. He cannot afford hospital bills.

HEALTH-CARE INSURANCE

Health insurance began in the nineteenth century, and ever since then it has been associated with wage laborers. In the nineteenth century, workers began organizing to contribute some of their wages to health insurance funds. This pattern has continued in industrialized countries for more than a century.

Ethical Issues Related to Health Insurance

The two main ethical ideas that are associated with health insurance are social solidarity (people standing together to help each other) and social justice (people working toward equality and fairness in society).

Health insurance and social solidarity. In a health-insurance agreement, people who face the same risks band together. Each person contributes an amount of money to a common fund. It is impossible to tell who may become unable to work, but in any group, it is certain that this will happen to some. So it is best to have a big group of mostly healthy people. If most members are healthy, it is likely that they will stay that way and that they will keep contributing to the plan without taking any money out. In this way, the small number of ill or injured people can be cared for. The healthy people are in solidarity with the injured or ill ones.

If the group is very small, it will not have as much money available, and even if only a few people develop costly problems, this could overwhelm the group beyond what it can afford. If a group has a large number of unhealthy people, each member will have to contribute more money so that all the unhealthy people can receive care.

In some societies, people believe that the government should provide health insurance and a certain basic standard of health care for everyone. This is true of countries in western Europe. In the United States, most people believe that health insurance should be provided by private companies who compete on an open market, just like other businesses. This view is controversial, but so far it has not been seriously challenged.

Kaiser-Permanente was the first Health Maintenance Organization (HMO) in the United States. Centered in California and started after World War II, Kaiser-Permanente is a prepaid plan. This means that patients pay a set annual fee to the HMO, which provides them with all needed care. Since the HMO agrees to this set annual fee, it must make efforts to reduce costs. The fewer expensive medical treatments, the more likely that the HMO will succeed financially. HMOs are supposed to provide care more equitably and affordably. Kaiser-Permanente provided the site for research indicating that cesarean section births were much overused and that many benefits are attached to natural birth, which is also much cheaper.

In 1947, the United States signed the Universal Declaration of Human Rights, which, in Article 25, states that "medical care" is a fundamental human right. At the end of the century, medical care has yet to be recognized as a fundamental right in America.

Health insurance and social justice. The concept of social justice is the second main ethical idea associated with health insurance. Social justice involves concepts of equality and fairness. Is it fair for some people to have health insurance and access to medical care, while others do not?

Originally, people thought that only wage laborers needed health insurance because other people were less vulnerable to losing work and money when they became ill or injured. Health insurance was thought of as a kind of social justice for poor, working-class people. Also, if working-class people had insurance, they would be more secure, and this would ultimately help their bosses and the entire system.

In time, workers began to expect health insurance as part of their jobs. In the 1940s, wages and prices were controlled, so people could not expect to receive raises. Instead, workers bargained for other benefits such as insurance.

Health insurance and human rights. Do we have a basic human right to receive health care? Some people believe that we do and that, as a part of social justice, everyone should contribute to ensure that the poor, as well as the rich, get care when they need it. People who believe health care is a basic human right say that it is so because it meets human needs. It provides relief from suffering, prevents unnecessary death, and helps people live and work well.

Discussions of human rights were not always a part of people's understanding of health care. Originally, people only wanted to protect the working class from the effects of major illness. Since then, the idea of health insurance has expanded and grown.

In the United States, there are no laws stating that everyone must have health insurance or that everyone has a right to health care. People who believe in these rights push to encourage individuals and lawmakers to help others who need care but cannot afford it. These people are not aiming to get private insurance for everyone. They want social change so that everyone in society will feel solidarity, contribute, and have access to health care.

Organizing and Financing Health Insurance

The basic idea behind any kind of health insurance is risk sharing. This means that people who face the same potential dangers (car accident, fire damage, treatment costs for illness or injury) band together and share their risks. These people all pay a regular amount of money, called a premium, to an insurance company. This company promises to pay for health care for any member who needs it. Even if a member does not need immediate health care, he or she has the security of knowing that if something bad does happen, it will be paid for.

This risk sharing among members of a group may bring up different ethical questions. These questions are shaped by how the

> "No other industrial nation in the world leaves its citizens in fear of financial ruin because of illness."
>
> —Edward M. Kennedy, U.S. Senator

insurance company organizes and finances the common fund used to pay for the members' health care. For example, a group might be organized to include older people, people who all work at the same business, or an entire labor union. The common fund might grow by charging everyone the same amount. Or the company may charge more money to people with higher health risks, or to people of a certain age group, gender, or ethnic group. Is it ethical to charge some people more? Is it ethical to base these increased charges on age, gender, or race?

The major development in U.S. health insurance in the 1930s and 1940s was not led by government or business. It was led by nonprofit corporations such as Blue Cross, which provided payment for hospital services, and Blue Shield, which provided payment for doctors' services. Because these companies were nonprofit, they gave U.S. health insurance a strong feeling of social solidarity.

The United States used these plans as a base for its own social insurance plans. Six aspects are typical of U.S. plans: (1) leadership by nonprofit corporations; (2) a shift from systems that charge everyone the same amount to systems that charge more or less based on the individual's health or other factors; (3) people's desire for insurance that covers many different health needs; (4) payment for basic services and injury compensation; (5) insurance companies' preference to provide services to groups of people instead of to individuals; and (6) the public's inability to decide whether insurance should be provided by private companies or the government.

Conclusion

Finding a way to provide as good a care as possible to the largest number of people has been a concern of Americans for a long time. In a 1946 message to the California legislature, Earl Warren (1891–1974)—Chief Justice of the U.S. Supreme Court and Governor of California—clarified not only the issues to be solved, but the intent of his administration's proposed insurance bill:

"We do not want to put the medical profession on the public payroll, nor do we want to deprive the individual of the right to select his own physician. . . . Our major purpose is to spread the cost of medical care among all the people of the state."

In the 1990s, consumer groups, business leaders, politicians, and health-care workers began to demand a change in the U.S. health-insurance system. In all industrial nations, health care was becoming so expensive that it was difficult for anyone, even insurance companies, to afford it.

In the United States, business, government, and consumer groups demanded that health-care expenses be controlled. Some wanted to do this by making health care a competitive consumer market. They believed that if health-care providers had to compete for customers, they would work hard to provide the best care for the least money. Others believed the government should step in to put limits on health-care budgets and regulate the way services were provided. Both groups agreed that health-insurance providers played a big part in the spending frenzy and that the companies had to stop this uncontrolled spending.

Controlling health-care expenses affects the social solidarity that is part of health-insurance plans. The governments of many European nations and Canada have tried to control their total expenses without cutting health care to their citizens. In the United

> "Health insurance has become a cata-
> clysm that threatens to bankrupt the
> family budget, the corporate budget,
> and the national budget."
> —John D. Rockefeller IV (1990)

States, social reformers, the public, and health-care providers began to demand a change in the U.S. health-care system. They wanted a universal system of health care that would provide a basic standard for everyone. These reformers described current health-care systems as unfair because many poor people could not afford them. During a thirty-two-month period in 1990–1992, one-fourth of all the people in the country were without health insurance for at least one month. By comparison, more than one-third of all African Americans and almost one-half of the Hispanic population did not have insurance.

The public became aware that more and more people did not have insurance. Even middle-class people were beginning to lose the insurance formerly provided by their jobs. More people became unhappy with the current situation and demanded changes in the entire health-care system. In the late 1980s and early 1990s, dozens of proposals were made, suggesting ways in which these changes could take place. These proposals asked several key questions. Should America continue its combination of government-provided insurance and private insurance—or should it begin a system where there was only one insurance provider? Should this one provider be the federal government—or should each state provide insurance? If there were to be many insurance providers, who would make sure that every citizen was covered?

Nevertheless, at the close of the century, no progress had been made toward a universal health-insurance system based on a right to health care.

HEALTH-CARE SYSTEMS

We define a health-care system as the way in which health care is provided and paid for. It includes decision making about who will get the care and who will receive what services. Also included is the structure that determines how much money to spend and who will provide the care with what equipment in what hospitals and other facilities.

The goal of a health-care system is to improve the health of the general population in the best possible way, depending on a society's resources and needs. During the twentieth century, most countries and the United Nations agreed that access to health care should be a basic human right. This means that they believe everyone should have health care when it is needed. If we look at different health-care systems, we must consider how well each one lives up to commonly held values.

The size and organization of a particular system depend on many factors, including the unique culture and history of a population or country. Considering what kinds of services are "health care" may differ from country to country, depending on how

technologically developed the country is. Some countries emphasize preventing disease, while others provide treatment only after people have become ill. Different cultures have differing ideas about what health is, what disease is, and who should be able to provide care.

An inventory of different views on health-care rights:

1. All citizens have a right to needed health care.

2. Health care is an earned commodity that only those who can pay for should have.

3. People who do not take good care of themselves have no right to health care.

4. Each citizen has a right to medical care according to real need, not mere want, consistent with efficiency and with what society can afford.

Different ethical values can also influence health-care systems. These values include respect for patients' and providers' rights to make their own decisions about health care. They also include the ideas that people should get the best health care possible for the lowest cost and that health care should be provided fairly to everyone. Balancing these values in health care has been a problem in the United States. Public opinion polls show that most Americans believe that access to health care is a basic human right. Yet most Americans also believe that people should be responsible for themselves and not rely on the government for care. At the same time, many Americans fear government interference in people's private

A doctor weighs a patient at his street clinic in Calcutta, India (1989).

lives and believe that a national health-care system would lead to this kind of interference.

Another influence on the structure of a health-care system is how much money a country has to spend on it. If a country is wealthy, it tends to spend more on health care. Even though everyone values health care, if a country does not have much money, it may spend what it has on providing food, shelter, and military defense instead. The amount of money a country has usually determines how much it spends on health care, but it does not determine what kind of health-care system the country has. Poor countries have as many different kinds of health-care systems as wealthy countries do.

Public Versus Private Control

In developed countries other than the United States, a government-run program dominates the health-care system. Often these countries also have a small but important private market for health care. This means that you can go to the government-run system for care or, if you have the money, to a private doctor.

In the United States, the government does not take great responsibility for health care. Private insurers, employers, and other providers pay for over 60 percent of health care. The United States spends about 14 percent of its **gross national product** (GNP) on health care. This is a higher percentage than any other nation's. (Canada, the next-highest spender, spends 10 percent of its GNP.) But even though the United States spends this money for health care, the government pays for only 40 percent of all health care. This is the lowest level of government health-care spending among all developed countries. The government of Austria, the next-lowest spender, pays for 68 percent of all health care in that country.

The government health-care systems of developed countries can be divided into two categories. Some have complete health-care programs and strong government control of almost all aspects of health care. In these countries, the government manages care payment and how care is provided, and makes sure it meets certain standards of quality. These countries include Great Britain, the Scandinavian countries, and the countries of the former Soviet Union. In other countries, the government role is limited to paying for or guaranteeing health insurance for all citizens. These countries include Germany, Belgium, France, and Canada.

Both types of systems involve government payment for care or laws that guarantee that everyone will have care. In both systems, the government pays health-care providers for their services and sets policies that regulate where care is provided and who provides the care. Both the ability of patients to decide where they get health insurance or health care and the ability of health-care providers to decide how much they will charge for their services are limited by

gross national product the total value of all goods and services produced in a nation during a specified time period

As early as 1910, the American Medical Association (AMA) was interested in a national health-insurance program. In 1916, it drafted legislation for this purpose. But soon afterward, the AMA was influenced by a different group of doctors who completely opposed federal involvement in medical-care delivery. Since that time, the AMA has consistently opposed any notion of a right to health care.

people's sense of responsibility to the community and by their relatively high trust in government.

Financing

How a society pays for health care shows what it thinks is valuable and important. As noted above, the United States is the only developed nation where most health care is not paid for by the government. South Africa, the Philippines, and Indonesia are examples of underdeveloped countries where the government does not pay much for individual health care. Certain treatments can be very expensive, and some people's care is much more expensive than others'. If a country does not have a broad-based government system for paying for everyone's health care, then people will get care only if they are able to pay for it. In effect, cost determines rationing of health care and only relatively wealthy people get all the care they need.

Government's share. In 1883, Germany was the first country to create a government-controlled system of paying for personal health care. Since then, most economically developed countries have created their own government health-care systems. In Great Britain and the former Soviet Union, the national government pays for almost all health care. In Canada, both national and state monies pay for health care. In Germany, France, Belgium, and the Netherlands, workers contribute taxes from their wages to a fund that provides health insurance to everyone.

All countries have an unofficial private market for health care that is not part of the government system. This private health care makes services available to people who are rich or politically powerful. Countries that have strong central control of health care tend

▶ Secretary of Health John Gardner of the Johnson administration outlines the implementation of Medicare, a program signed into law by President Lyndon B. Johnson on July 30, 1965.

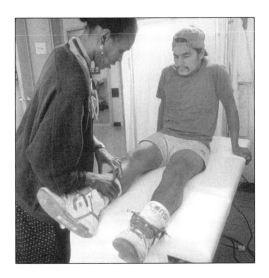

A physical examination at the Pine Ridge Indian Hospital (South Dakota). Federal money pays for the health-care system of Native Americans.

to have a small private market, while some countries with mixed systems, such as Japan and Australia, have a larger number of people with private health insurance.

The percentage of government payment for health care in the United States began to increase in the twentieth century. Between 1950 and 1993, the amount jumped from less than 10 percent to nearly 40 percent. However, a key part of health care that is still missing is a government-guaranteed system of health insurance for everyone. Most people receive health care through insurance provided by their employer, or they pay for the insurance with their own money. There are also a variety of health-care programs in the United States that are paid for by federal and local governments. The payment for these programs comes from federal and local income and employment taxes. Some states use money they make from selling lottery tickets.

One of the largest unified health-care systems in the United States is the one providing care for people who are on active duty in the military. This system is similar to the centrally controlled systems of Great Britain and the former Soviet Union in that money from the federal income tax pays for it. Money from federal taxes also pays for the health-care system for veterans and Native Americans. The Medicare program, which provides care to elderly retired people and their dependents, is mostly paid for with taxes on working people's wages. A combination of state and federal tax money finances the Medicaid system, which provides care for low-income people. Some poor people who are not eligible for Medicaid obtain care at city or county public hospitals. Their care is paid for by state and local taxes.

Evolution of the health care system. The fact that the U.S. government does not pay for most of its citizens' care reflects American values and is a result of several historical events. Before 1900, a few health-insurance plans were based on employment, but this kind of insurance did not really spread until World War II. During the war, the government put a "freeze" on people's wages in an attempt to prevent prices from going up. But the government did not freeze other benefits to employees, and employers began providing health insurance instead of raising wages. This rapid spread of insurance had begun in the 1930s, when hospitals created Blue Cross insurance programs to help pay for hospital care. Doctors followed this lead and created Blue Shield programs, which provided payment for their services.

As more companies began to offer health insurance, private insurance companies realized they could expand their markets, and they encouraged people who already had health insurance to buy other kinds of insurance, such as home and car insurance. The market also grew when the federal government decided not to tax health-care benefits that people received from their employers. The United States has a vast number of insurance companies, each with

its own marketing, packages of health services, monthly costs, percentages that the patient must pay, billing, and reimbursement methods. There are thousands of private doctors, clinics, and hospitals. All these different systems and care providers have created a huge, complicated bureaucracy that is estimated to cost over $50 billion per year.

Access and Delivery

Access defines how people get health care and what services are available. There are several basic approaches to access. One is provided by systems with strong central control—Great Britain, the Scandinavian countries, and the countries of the former Soviet Union. In these countries, most health-care providers are paid employees of the government. These countries offer one set of health-care services as defined by the government. They emphasize basic care by general practitioners who treat a variety of health problems. These systems tend to tightly control the number of doctors and hospitals that provide expensive technical care. In some countries, this tight government control results in patients having to wait a long time for some services, and it limits access to advanced technologies. These strongly controlled programs may intend to provide equal care for everyone, but because of these limits, some people might not get lifesaving care when they need it.

Countries with less centralized systems vary more in the level of access and in the way of providing it. In some countries, poor people do not have as much access to health care as do employed people. When the government does not pay for the full cost of treatment, some doctors may refuse to treat poor people because they know they cannot pay.

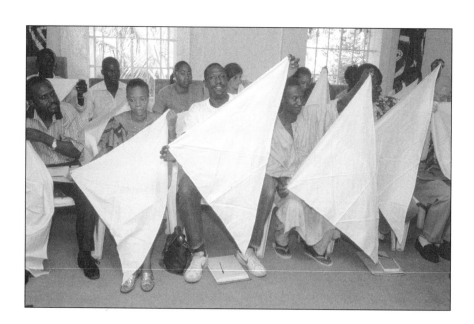

▶ Villagers learn how to fold bandages at a Red Cross first-aid course in Gambia.

The American College of Physicians made a major proposal for health-care reform in 1990:

All Americans should be able to obtain appropriate health care services, irrespective of age, race, sex, financial status, or place of residence. We believe that assuring access to health care services would achieve improvements in health status, decrease incidence of morbidity [illness] and mortality, and, possibly, contribute to an increased life-expectancy among those groups with the least access to care.

In other countries, such as the United States, even if people have insurance, they often have to pay some of the bill themselves. People who have low-paying jobs may not be able to afford to do this. Also, because the government does not control where doctors live or where hospitals are located, some areas have many hospitals while others have none. Or an area may have doctors but no heart specialists, which creates a problem for patients with heart disease.

This wide variation in access to care is very obvious in the United States. More than 35 million Americans do not have health insurance. An estimated 20 million more have insufficient insurance. Studies show that people who are poor and have no health insurance do not get proper health care—even though they may need more care.

Uninsured people might be able to get high-technology care if they live in a city where there is a public hospital. They also have access to emergency rooms, but not to routine care. Research shows that these people do not improve as much as other people do from the treatment they get and that hospitals do not use high-technology care as often with patients who are not insured. There is also growing evidence that if people do not get basic care, their overall health will be worse and their care will eventually cost more than it would have if they had received basic care in time.

Reimbursement

Health-care providers receive payment for their services by the government, insurance company, or other payer through reimbursement. How much providers are paid, and how they are paid, has a major effect on access, costs, and quality of care. In some countries, such as the United States, Canada, France, and Belgium, doctors are paid for each service they provide. Doctors bargain with insurers or the government to determine how much they should receive. There is often an agreement that, in certain cases, doctors can charge patients more than the allowed price.

Pay-per-service. Some people fear that when doctors are paid for every service they provide, they might perform more services than the patient needs, just so they can get more money. In this system, however, doctors keep their independence from the government, which some people believe is a good idea. Doctors are encouraged to work hard because they are paid according to how much work they do. And when doctors are paid for providing care, they have no incentive not to give patients the care they need. Hospitals and nursing homes are paid also according to the services they provide. The risks and benefits for these institutions are similar to those for individual doctors.

Capitation. A growing number of insurers in the United States has switched to new payment methods. One method, called capitation, involves paying health-care providers a set amount for

each person's care each year. Another method involves a set payment for each case, which, because it involves a set amount of money, forces providers to limit the amount of services they provide. It also allows them to determine how they will spend the money and what services they will provide.

At the same time, case-based payment and capitation create a conflict between the interests of the health-care provider and those of the patient. What happens when the patient needs care that is more expensive than the set amount? The Rand Health Insurance Study in the 1970s looked at health-maintenance organizations in these kinds of systems. The study found that the poor in these systems were getting a lower quality of care than the nonpoor. The systems were giving poor people less care and then using the money they had saved to give better care to others.

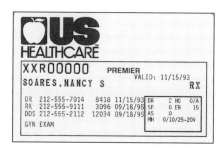

▶ Two examples of health maintenance forms: an ID card for an insurance company (left) and a Health History Summary form.

"Cost containment too often becomes health care containment, and it affects the least powerful population sector with the quietest voice."
— A physician at a conference on health care

Costs and Cost Controls

Since 1960, in almost every country in the world, health-care expenses have been increasing almost twice as fast as other national expenses. Critics are concerned that governments spend so much on medical care that there is less money for other goods and services. This is especially true in the United States. Even though the United States is a wealthy country and spends more than any other country on health care, not everyone has access to health care. Critics are also concerned about other social problems like bad schools, homelessness, poverty, and crime. In addition, there is evidence that more health care is not necessarily better. And many medical-care services may not actually be as helpful as they seem. In developed countries, people spend a great deal of money on such care. In poorer countries, governments do not even have the money to provide basic public-health services like immunization and sanitation.

In each country, the way in which the health-care system reacts to the problem of increasing costs shows how the country is organized and what its people value. In countries with strong central control, there is more pressure to prevent increased spending and

to control the use of high-technology health services. These countries believe that everyone should have access to health care, but people who need expensive, high-technology care to save their lives may not get it.

In the United States, there is less interest in providing services to everyone or in preventing spending from increasing. Some people have tried to reduce health-care costs by limiting how much people use the services. Others believe that if health-care providers compete with each other on an open market, this competition for customers will keep costs low. Some reviews of care provision have made care more efficient, so that costs may be temporarily reduced, and patients spend less time in the hospital. But these changes have not been long-lasting. Health care is still rapidly becoming more and more expensive.

Because of this rise in costs, employers have been forcing employees to pay for more of their own care. They are increasing the amount employees must pay for health insurance, they are not paying for health care for employees' families, or they refuse to insure people at all. Some private insurance companies refuse to insure people who are likely to need care, or they base the amount people pay on how healthy they were in the previous year.

These problems with insurance companies, and the increased number of people who work for small businesses that do not provide health insurance, have increased the number of people in the United States who do not have any kind of health insurance.

> When you have 25 percent of the city not being able to afford care, health becomes everyone's responsibility. . . . What is the moral responsibility of a hospital? Do you assure basic services for the many, or do you retain complex sophisticated services that tend to affect the relatively few?
>
> —Robert D. Grumbs, director, New York City Health System Agency. Quoted in *The New York Times*, March 12, 1989.

Choices for the Future

Every country will have to face the difficult task of controlling health-care expenses for itself. Governments may have to limit the use of expensive, high-technology treatments that provide small benefits to very few people at a great cost to the community.

This problem brings up the tension between people's right to make their own decisions and to get the care they need, and concern for the good of the community and society. Suppose someone has a disease that cannot be cured but requires very expensive, high-technology treatment just to keep the patient alive? Should that person receive treatment even if the same amount of money could provide basic care to prevent disease in the entire community? The drive to provide everyone with equal care conflicts with the real limitations of money and resources. The system cannot afford to give everyone expensive, high-technology care. But refusing care to people can be seen as an abuse of their personal freedom and their right to the care they need.

Balancing these conflicts will be especially difficult in the United States, where there are so many health-care systems and where people do not trust government involvement in health issues.

MEDICAID

Before the Medicaid program began in 1965, the American medical system did not provide care for people who could not afford to pay. If poor people needed care, they had to rely on charity and public hospitals and clinics, along with limited help that was based on public welfare. Also at that time, African Americans were segregated and had even more trouble than the poor in getting care.

In the 1950s and 1960s, Americans became concerned about people who did not have health insurance, and in 1965 Medicaid was passed. Medicaid provided medical insurance for poor Americans.

Initial Goals and Early Challenges

Medicaid is a program of health insurance for people whose income and personal possessions are below a certain dollar amount. This amount is determined by each state. The program was not designed to cover all poor people but focuses on people who receive certain kinds of welfare payments. Medicaid mainly helps children and single-parent families who receive Aid to Families with Dependent Children, and elderly, blind, and disabled individuals who receive Supplemental Security Income. Unmarried individuals and childless couples cannot be in these programs. Neither can two-parent families, no matter how poor they are. These programs leave out almost half of all poor people who are not elderly.

At first, Medicaid was supposed to make sure that people received basic care and did not have to pay as much for medical costs. This meant that the poor and the nonpoor should receive the same services, and that the poor should go to the same hospitals as everyone else.

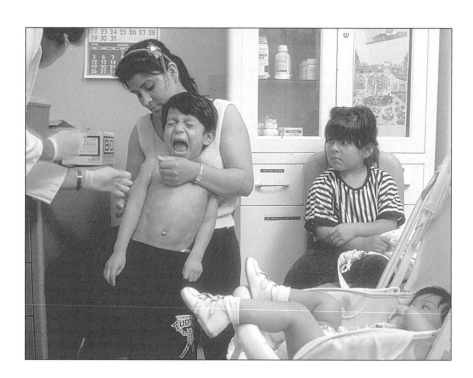

Children are being vaccinated in a public health clinic in California. Medicaid has made a difference in the health of poor people.

States pay up to half of the costs of Medicaid. The federal government pays the rest. The federal government sets guidelines, but the states determine who gets Medicaid, what services will be covered, how much health-care providers are paid, and how the program will be run. Although Medicaid programs all have a similar structure, Medicaid is actually fifty different programs, one in each state.

Only a few years after Medicaid began, federal and state officials became concerned about the program's rising expenses. The public became more concerned about cutting costs than about providing equal care to everyone. Between 1975 and 1990, the number of people receiving Medicaid grew slowly, from 22 million to 26 million. But government costs for the program grew quickly, from $12 billion to $72 billion.

Medicaid has never been a popular program. It is connected to welfare programs, which some people think waste money. Also, it is expensive, the people it benefits are poor, and half of them are children. Neither of these groups has much political power.

> Should physicians be allowed to practice where they want? Nationwide, the patient-to-physician ratio is about 500 to 1; in poor inner-city neighborhoods, and in some poor rural areas, it is about 2,000 to 1.

Impact on America's Health

It is hard to tell how Medicaid has really affected the health of poor people. It is clear that since Medicaid, the health of poor people has greatly improved. The rate of infants and mothers who die has been reduced, and death rates for diseases that can be treated are lower. The difference in health between African Americans and white Americans is not as great as before.

It is clear that if someone does not have Medicaid coverage or other health insurance, his or her health is affected. When Medicaid was taken away from chronically ill and poor adults in California, they had a harder time getting the care they needed and their health became worse.

MEDICARE

In 1966, when the Medicare program began, it was thought of as a way to provide medical insurance to elderly people who otherwise might not be able to afford it. But over the next twenty-five years, people began to see Medicare as a tool to influence the costs and actions of doctors, hospitals, and health insurance. This was because the Medicare system was so big, and because if Medicare refused to pay for a treatment, doctors and hospitals generally did not want to do it. In 1986, when Medicare was twenty years old, it gave itself a big birthday party. But in 1991, when Medicare was twenty-five years old, it was having serious financial trouble, and its birthday was not celebrated.

The Origins of Medicare

Medicare grew out of the political trends of the early 1960s. In 1964, the Democrats wanted an insurance program that would pay for elderly people's hospital costs. The Democrats had won heavily in the elections that year, and people believed this program was guaranteed. President Lyndon B. Johnson talked about this program during his campaign, and the new Congress of 1965 was on his side.

Because people believed the program would pass no matter what happened, they did not bother to discuss whether it would really be a good idea. Instead, they only talked about what it would do.

Implementing Medicare. Between 1966 and 1972, the government did not do much to regulate Medicare. Spending rose rapidly and the program did not develop good cost controls. From 1972 to the beginning of the 1980s, the American economy was troubled by a mix of inflation and unemployment. People talked about changing the American medical system. They began to think that medicine should be more like a big business, competing for customers. During this time, government regulation spread among federal and state agencies. The public began to demand reforms that would change the health-care system, bring costs down, and reduce government interference.

An example of the problems people had during this time is a Medicare program that paid for treatment for anyone who had kidney failure, not just for the elderly. This program began in 1972, and people were happy about it at first. But the program grew too quickly. More people were enrolled, costs rose, and the program became so complicated that people pointed to it as a symbol of what was wrong with Medicare and other government health-insurance systems.

Reforming Medicare. Throughout the 1970s, experts suggested many different ideas for change in the health-care system. But changes in Medicare were the responsibility of government policy specialists, congressional committees, and the agencies that ran the program.

In the 1980s and 1990s, several developments took place. First, people tried to reduce Medicare spending. Initially, the elderly had to pay some of the costs of their care, and then hospitals and doctors were burdened with these costs. In 1983, a new method of paying for treatments began. Each health problem and treatment was listed and given a code, and a set amount of money was paid for it. If two people had the same kind of cancer and needed the same treatment, they received the same code, and the same amount of money would be provided for their care. This diagnosis-related group (DRG) method was developed by experts in hospital technology and bureaucrats, and it dominated work in hospitals during this

▶ Demonstrators in New York express their support for the Medicare program.

In a 1965 message to Congress, President Lyndon B. Johnson explained his determination to see his health-care bill passed in these words: "I am proposing that every person over sixty-five years of age be spared the darkness of sickness without hope."

time. If something did not fit within the codes, it would not be paid for. People believed that this scientific, nonpolitical system would solve Medicare's spending problems. A new government commission was formed to oversee the use of DRGs, and later in the 1980s another commission was formed to oversee payments for another part of Medicare, called Medicare Part B. These commissions were expected to keep costs under control but make sure that the quality of care did not change.

Twenty-five years after Medicare began, Americans were still upset about government spending and national budget programs. Medicare had a bad name, and some Americans felt that the elderly were unfairly using up all the money, leaving none for programs for children and for the future. In 1992 and afterward, policymakers continued to try to cut Medicare's spending, and Americans continued to debate about changing the entire health-care system.

Mental Health

ABUSES OF PSYCHIATRY

psychiatry a branch of medicine that deals with the treatment and prevention of mental illness

psychiatrist a doctor who treats patients who suffer from mental illness

The concept of the abuse of **psychiatry** brings to mind a situation in which a **psychiatrist** acts improperly, causing a patient to experience some sort of harm. But psychiatric abuse is much more complex than it appears to be at first glance.

Definitions

Psychiatric "abuse" is different from other undesirable practices that are best termed "malpractice." "Abuse" refers to the intentional, improper use of the knowledge, skills, and technology of psychiatry for a purpose other than helping the patient, or to harm individuals who should not be considered mental patients. Abuse by psychiatrists (or other mental-health workers) is usually done in collaboration with other people or agencies, such as a secret police or political authority.

Abuse can also be carried out by a psychiatrist to help someone exploit psychiatry for nonmedical purposes. Consider this example: A husband who knows that his wife is not mentally ill persuades a psychiatrist to commit her to a mental hospital. The husband is not concerned with the welfare of his wife. He wants to exert power over her, so he gets the psychiatrist to act as his accomplice.

The difference between "malpractice" and "abuse" depends on the psychiatrist's intent. Since the term "malpractice" is used in many different ways, the term "inadequate practice" comes closest to the meaning intended here. A psychiatrist who does not set out to use his or her knowledge, skills, or technology improperly but who uses them in an inadequate or unskilled way is engaging in malpractice. A case in point is the common practice of prescribing tranquilizers for a patient at the request of nursing staff. The nurses insist that they cannot otherwise manage the patient, even though the patient does not need tranquilizing drugs. In prescribing the drugs for the patient, the psychiatrist fails to adhere to a basic standard of practice that requires prescription of drugs only when needed by the patient.

Malpractice is different from "errors in clinical judgment" made in good faith. Psychiatrists, like any other professionals, make mistakes on occasion. Although the consequences may be similar, malpractice has not actually been committed.

Historical Background

In the latter half of the twentieth century, evidence emerged of such practices as the abuse of psychiatry for political purposes. Examples of psychiatric abuse to stop political dissent or disagreement were found in the former Soviet Union, as well as in Cuba. In

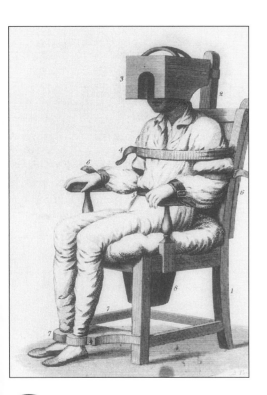

Before medication was available to calm a patient down, tranquilizing devices such as Benjamin Rush's chair (c. 1811) were used in mental institutions.

addition, psychiatric abuse was used in the torture and questioning of prisoners in Northern Ireland in 1971. The abuse of psychiatry by the Nazis during World War II, especially the program in which tens of thousands of psychiatric and mentally retarded patients were killed or sterilized without their consent, is the starkest instance of abuse in recorded history.

▶ A prisoner at the Vacaville penitentiary in California is being prepared for a lobotomy (1961). At the time, many psychiatrists believed that "criminality" was lodged in certain areas of the brain, and lobotomies on prisoners were a frequent occurrence.

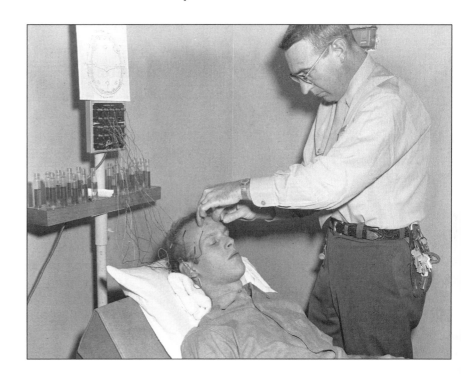

Lobotomy

In 1936, psychiatrists Walter Freeman and James Watts introduced the precision-lobotomy technique into American psychiatry. A narrow knife was inserted through a hole in the side of the skull and with a single motion the frontal lobe was severed. The frontal lobe section of the brain is responsible for emotion and for much of human behavior. As a result of this surgery, the patient would become calm, passive, and expressionless. But Freeman and Watts ignored the adverse side effects, including general intellectual decline and general to extreme lack of emotional responsiveness. Regrettably, by 1955, an estimated seventy thousand lobotomies had been performed in the United States alone. Later, Freeman introduced the infamous "ice-pick lobotomy," in which an ice-pick-like instrument was inserted above the eye, hammered through the eye socket, and then used to sever the frontal brain lobes. Many thousands of these surgeries were performed, predominantly on women. After the public and eventually psychiatry became properly critical of lobotomies, they were stopped. In the late twentieth century, only a few lobotomies per year were being performed in the United States, mostly on the criminally violent and with their full consent. Even this very limited use remained highly controversial.

 Ezra Pound returned to the United States after World War II to face charges of treason. Pound was found incompetent to stand trial and released.

see also

Mental Health: Commitment to Mental Institutions

Abuses in the United States. The abuse of psychiatry for political purposes in the United States has occurred occasionally. Two well-known examples are the famous poet Ezra Pound and General Edwin Walker. In the case of Ezra Pound, psychiatry was called upon to deal with a politically sensitive and difficult situation. Pound, indicted for treason following his pro-German and pro-Italian broadcasts in Italy during World War II, was facing possible execution. Although the evidence was unclear, Pound was thought to be incompetent to stand trial. As a result, he was transferred to St. Elizabeth's Psychiatric Hospital in Washington, D.C., where he spent thirteen years. The indictment against Pound was dismissed, and the poet was released. Whether psychiatry was used to free the U.S. government from an embarrassing situation or Pound was genuinely mentally ill remains a controversial question. His case clearly shows how psychiatry can be used by political forces.

The situation was similar in the case of Edwin Walker. Walker was a highly decorated major general in the U.S. Army who strongly supported the extreme right-wing position during the desegregation movement of the 1950s and 1960s in the South. He believed that the South should not permit African Americans to be integrated into society. Walker's mental competence came into question after he had been charged with a number of offenses related to his political activities. Although he was declared competent to stand trial (the case was later dismissed for technical reasons), the evidence in the case raises the possibility that the government used psychiatry to deal with a "troublemaker" more conveniently than through the legal process.

Abuses against women. This brief historical background should also mention the criticism of psychiatry as biased against women, who prior to the 1880s could be commited to a mental institution merely for publicly disagreeing with their husbands. The dramatic case of Mrs. Packard in 1860 shows how prejudice can destroy sound judgment. Her husband, a fundamentalist clergyman, insisted that she had dangerous religious beliefs. As a result, Mrs. Packard was committed without her consent to an Illinois mental hospital, where she remained confined for three years. Upon her release, she began a campaign against allowing the expression of opinions to be grounds for commitment to a mental institution.

The Vulnerability of Psychiatry to Abuse

Abuse appears to be more common in psychiatry than in other fields of medicine. This is probably because psychiatry is more vulnerable to it. This is the case in at least three respects: (1) the boundaries of psychiatry are blurred and poorly defined; (2) psychiatrists often make diagnoses without objective standards; and (3) psychiatrists hold enormous power over the fate of other people, even to the extent of committing them to a mental institution without their consent.

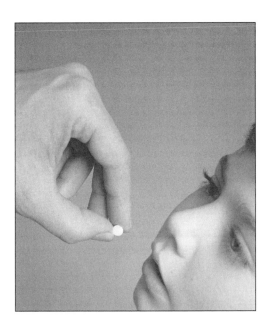

A boy is being given a Ritalin pill. Critics claim that the drug is often used to reduce a high but healthy level of energy in active children.

Ritalin is a drug used widely to treat hyperactive children. Parents consent to Ritalin for their children, sometimes at the recommendation of schoolteachers. Critics claim that this drug is being used as a substitute for discipline, parental time and energy, and environments that allow children to be as active as they desire.

Blurred boundaries. Since the 1960s, psychiatrists have debated among themselves what their legitimate role is. Attitudes vary to the point of contradicting one another. The following views, expressed by former presidents of the American Psychiatric Association, reflect these different attitudes. In 1969, Ewald Busse, then president of the association, argued for a limited role, in which the psychiatrist restricts his or her focus to the suffering patient in an effort "to reduce pain and discomfort." Dr. Busse's colleague, Raymond Waggoner, had a very different attitude about the role of psychiatrists. In 1970, Dr. Waggoner called upon psychiatrists to pursue basic "social goals" and to concern themselves with "individual liberty" and their responsibility to the community. In Dr. Waggoner's words, psychiatrists should be not only practical doctors but also "dreamers with a vision for the future."

Lack of objective standards in diagnosis. Although psychiatry has been a branch of medicine since the second half of the nineteenth century, the profession still wrestles with the basic question, What constitutes mental illness? No satisfactory standards exist to define most of the conditions psychiatry deals with. Despite the attempts of the American Psychiatric Association in 1987 to develop common standards, many psychiatric diagnoses come from clinical observation only and lack the objective tests used in other medical fields. Objective tests that determine the presence or absence of mental illness are rare.

Unlike other doctors, the psychiatrist must also rely on social standards and value judgments. Society would not be able to determine what is normal if it failed to label certain acts and certain people as abnormal or antisocial.

The psychiatrist's power. Even though the boundaries of psychiatry are poorly defined and the standards for diagnosis are vague, the psychiatrist is legally certified to commit a person suffering from mental illness to a hospital without the person's consent for at least a seventy-two-hour period if deemed necessary to protect that individual's welfare or the welfare of others. In this way, the psychiatrist is given tremendous authority. The person suffering from (or suspected of suffering from) mental illness is deprived of his or her freedom, loses many civil rights, and is subject to a wide range of hospital regulations.

Certain commitment laws have been subject to careful observation and improvement. But psychiatrists are often caught in the dilemma of having to make a judgment about a person's clinical needs while still protecting his or her civil rights. The civil libertarian would argue that the individual's right to freedom should be protected above all other considerations. Other people might argue that psychiatrists sometimes have an obligation to take necessary measures, including forced hospitalization, to protect the patient, society, or both from harm.

Soviet Psychiatric Abuse

Poorly defined boundaries, subjective standards in reaching a diagnosis, and the authority to hospitalize persons without their consent combine to make psychiatry especially open to abuse. The most clear-cut example was the way psychiatry was used in the former Soviet Union to squash political and religious dissent.

Boundaries set by the state. Soviet psychiatry's boundaries were drawn in a way that made the entire profession subject to the influence of the Soviet state and the Communist Party. A small group of psychiatrists, acting on behalf of the state, determined the role and function of all Soviet psychiatrists. Even if the boundaries of psychiatry had been clearer, the Soviet dictatorship ensured that psychiatrists could not function independently. Since the boundaries were blurred, it was easier for the state to exert control and mold the profession according to Communist Party beliefs. The Soviet state's position that the interests of society were at least as important as those of the individual led to the complete undermining of the respect for **self-determination** and independence.

self-determination the interest people have in making important decisions about their lives according to their own values

The Soviet case is a glaring reminder that psychiatrists may function in a state whose interests do not serve those of the public. It follows that psychiatrists must be above politics and act independently with regard to ethical standards.

Classifying mental illness. The lack of objective tests to diagnose mental illness permitted Soviet psychiatry between the 1950s and the 1980s to come up with its own system of classification. Professor Andrei Snezhnevsky quickly became the leader of the psychiatric establishment during the 1950s. From that powerful position, he launched a unique system of classifying mental illness. An important result was the shift in the way **schizophrenia**, usually considered a major mental disorder, was defined. Snezhnevsky claimed that since the illness could be present in a person showing only minor symptoms, schizophrenia was much more common than once thought. A particular form of the illness, "sluggish schizophrenia," named for its extremely slow rate of progress, accounted for the wider use of the diagnosis. From the 1960s on, a diagnosis of sluggish schizophrenia was often applied to people who disagreed with the political or religious policies of the Soviet Union and was used by the state to punish them.

schizophrenia a mental disorder characterized by a view of the world that does not match reality, sometimes causing odd behavior

Although the Soviet system of classifying mental illness was not originally set up with the purpose of stopping dissent, its vague ideas helped to label as mentally ill people whom psychiatrists elsewhere would have termed normal, mildly eccentric, or at worst **neurotic**

neurotic fearful or worried for no apparent reason

The misuse of psychiatry in the Soviet Union shows that psychiatrists must condemn abuse of their profession wherever it occurs. Even though such protests are not strictly medical activities, they point to the political and social role psychiatrists may be required to play.

Conclusion

The history of psychiatry has been tainted by the occurrence of abuse. The Nazi and Soviet cases are glaring examples of this abuse. Public attention to such cases has increased ethical sensitivity among psychiatrists. Although this awareness provides a safeguard against abuse, both the psychiatric profession and society need to defend themselves against any forces that would exploit psychiatry and jeopardize its integrity.

Related Literature

In Michael Crichton's *Terminal Man*, Harry Benson stops worrying about life and learns to love the electrical charge he gets from 40 electrodes planted in his brain.

◆◆◆

In John Updike's short story "The Fairy Godfathers" (1976), two lovers talk about their psychiatrists. Neither patient seems to have any selfhood or reality outside of what their doctors tell them, including what they tell them about their relationship as lovers. The two patients seem to be reduced to case material for their weekly appointments with their doctors. The psychiatrists have altogether too much power over their patients' lives.

COMMITMENT TO MENTAL INSTITUTIONS

psychiatric having to do with the treatment of mental illness

psychiatrist a doctor who treats patients who suffer from mental illness

ll over the world, it is legal to send mentally ill people to **psychiatric** hospitals even when they do not want to go. In the United States, this is sometimes done when a person is charged with a crime. A judge may rule that a person is too mentally ill to understand a criminal trial or to assist in defending him- or herself. Sometimes, a person may be tried for a crime and found not guilty by reason of insanity. In both cases, such individuals will be "committed" to an institution for mentally ill criminals.

Most often, no crime is involved when mentally ill people are committed to institutions. In this type of commitment, called civil commitment, a **psychiatrist** decides that the person must be sent to a mental hospital because of his or her mental condition, even if the person objects.

Committing people against their will raises major ethical issues. It involves depriving individuals of their freedom for days, weeks, or longer, usually by keeping them in a locked mental hospital. In most states, a person can be briefly (usually seventy-two hours) committed against his or her will on an "emergency" basis

Patients in an Ohio insane asylum, 1967.

Frances Farmer (1913–1970) was an excellent student at the University of Washington when she wrote an essay in a radical socialist magazine for which she was awarded a trip to the Soviet Union. In the 1930s, Frances began a promising acting career both in films and on Broadway. But her outspoken socialist views, her discontent with romantic roles, and some heavy drinking all created friction with her mother, who was able to have Frances involuntarily committed to a mental institution. Frances was angry about the legal power of psychiatrists and of her mother. Her resentment against their power aggravated the psychiatrists, resulting in both repeated electroconvulsive therapy and, eventually, a lobotomy. Farmer was held against her will for the better part of the 1940s. Finally released, she was never the same. Her autobiography, a classic indictment of psychiatry, entitled *Will There Really Be a Morning?*, was published posthumously in 1972. Her experience of psychiatric abuse was recreated in the 1981 film *Frances*.

simply if one doctor signs the right form. After the emergency commitment form is signed, the person is taken to the nearest locked mental institution. The medical staff of the institution usually has the right to decide whether the person should be committed or not. In most states, a hearing before a local judge is held within two to three working days, so that the judge can decide whether the person should continue to be held in the mental institution.

Most people who enter mental hospitals choose to admit themselves. A small number of voluntary admissions take place when the individuals are told that they will be committed against their will if they do not enter the hospital "voluntarily." Although these individuals would not have entered the hospital completely voluntarily, there seems nothing wrong with giving them a chance to avoid an involuntary commitment. But it would be unethical for a doctor to convince a person to enter a hospital by threatening a commitment that in fact would not be carried out.

Legal Standards for Commitment

Psychiatrists disagree about the justifications for a person to be legally committed. Requirements differ from state to state.

Some people think that doctors should be able to commit anyone they believe would benefit from commitment. At one time, many states' laws permitted this. Until 1981, people in Arizona could be committed if they were "mentally ill and in need of supervision, care or treatment." Most people believe these standards are too broad. Many individuals with depression could be described as being mentally ill and in need of treatment, but few people would agree that depressed people should be forced into a mental hospital if they do not wish to go.

"Serious disruption." Many psychiatrists believe that in order for a person to be committed, the person's mental illness should have caused a "serious disruption" to his or her functioning.

Dorothea Dix (1802–1887) advocated better treatment for the insane, persuading legislators to vote for the establishment of asylums. Dix first became aware of the poor treatment and squalid living conditions of insane and disturbed people when a young clergyman asked her to begin a Sunday School class in the East Cambridge House of Correction. Shocked to discover that insane people were thrown in with criminals, left unclothed, chained up, and otherwise mistreated, she traveled to institutions all over the state of Massachusetts and submitted a report of her findings to the state legislature.

Wanting to kill or seriously harm oneself or others is one kind of serious disruption, but not the only one. Certain individuals behave so strangely that even though their behavior is not immediately harmful, it may cause them serious social or financial harm in the long run. Under a "serious disruption" standard, many such individuals could be committed.

Harmful behavior. Some psychiatrists, and many people concerned with civil liberties, believe that the standards for commitment should be stricter. A person should not only be diagnosed with mental illness but must pose a serious physical threat to him- or herself or to others. Some in this group would also require evidence that the person has recently behaved in a physically harmful way, but most believe that threats of physical harm are enough. Most also think that a person who is a serious physical threat to him- or herself through self-neglect (such as starvation or not treating a serious illness) should not be committed unless the threat of danger exists in the near future.

Opposition to standards. Finally, there are those who believe that psychiatric commitment is never proper, and so there should be no commitment standards at all. Thomas Szasz, a psychiatrist, is the best-known supporter of this view. Szasz believes that "mental illness" is a made-up concept. He argues that people considered mentally ill should be judged only by the standards of criminal law. That is, if they have broken a law, they should be arrested, but otherwise they should be free. Szasz believes that strange or deviant behavior is not an illness, and that psychiatrists who commit individuals thereby become tools of society's narrow idea of normal behavior.

Most psychiatrists and civil libertarians disagree with Szasz's position for two reasons. First, most scholars believe that some psychological conditions do fall within the definition of "illness." Second, most believe that even if commitment forces a person do what is for his or her own good, it is sometimes the ethical thing to do.

Legal aspects of commitment. Most states have commitment laws that are not at either extreme. Some states advocate that people should be committed if their behavior shows a "serious disruption of functioning." Moreover, danger to oneself or others is only one of many aspects of serious mental illness. It is cruel and improper to fail to commit mentally ill individuals, often homeless and wandering the streets, when they would clearly benefit from treatment in an institution.

Other states have laws holding that people should be committed against their will only if they are dangerous to themselves or others, because commitment for other reasons is a danger to freedom and civil liberties. Doctors could misuse their power (and have) to commit people with unpopular political views. Also, in a free society people should be free to act a certain way even if other people find it self-defeating, and to reject opportunities for help.

bipolar disorder a mental illness characterized by alternating manic and depressive states; also known as manic-depressive disorder

paternalistic acting like a father or other authority figure

In 1949, journalist Albert Deutsch wrote *The Shame of the States*, in which he documented the appalling conditions of mental institutions. Throughout the 1960s and 1970s, many patients with mental illness were freed from mental institutions. Yet many of these patients did not benefit from life in the community, where their care was more difficult for families and professionals. As a result, people with severe mental illness make up about one-half of the homeless in the United States. A classic book about this problem is *Madness in the Streets: How Psychiatry and the Law Abandoned the Mentally Ill* (1990), by Rael Jean Isaac and Virginia C. Armat.

Kenneth Donaldson had been committed to a state mental hospital by his father after a judge found that he suffered from paranoid schizophrenia. He lived in the hospital for fifteen years without ever receiving treatment. Donaldson then sued the state for his having been held in involuntary confinement. The case ended up in the Supreme Court, which decided in 1975 that mentally ill persons cannot be denied their liberty as long as they do not pose a danger to themselves or to others.

▶ In 1975, the U.S. Supreme Court ruled that insane persons cannot be confined involuntarily if they do not pose a danger to others. Here Kenneth Donaldson, who brought the suit, holds up the Court's opinion.

Some mentally ill people could be committed under the broader set of rules but not the narrower one. People with **bipolar disorder** may, in one stage of illness, go on spending sprees and waste all their money. They may refuse all medical help while in this stage. Relatives and friends may believe that these spending sprees are part of the illness and that they should be stopped. However, these people are still not dangerous to themselves or others under the laws of many states.

Conceptual Issues Underlying Commitment

When discussing whether commitment is right or moral, one must decide whether the commitment is intended mainly to help the mentally ill person or to help others whom the person may be endangering. This distinction is not always clear, because it is usually to the advantage of the mentally ill to be kept from harming others. The harm they might cause others could be a serious crime. The mentally ill individuals could then be put in prison or in a mental institution for a long time. The distinction should nevertheless be made between committing people to help *them* and committing people to help *others*.

Commitment as protection. To the extent that commitment is made to help the person committed, it is almost always a **paternalistic** action. That is, the commitment is intended to help the person but deprives the person of freedom and is done without the person's consent.

When commitment is not made to help the mentally ill person (in other words, is not paternalistic), it must be ethically justified on other grounds. Commitment laws in the United States usually do not allow mentally ill individuals to be committed simply to try to prevent them from harming others, although actual harmfulness to others does permit commitment.

Predicting possible harm. The following factors are used in predicting possible future harm when deciding whether to commit someone: (1) the *criterion* (standard) is the outcome that is being predicted (say, a person's suicide); (2) the *cues* are small, separate pieces of information about a particular case (a person's age, sex, or history of strange behavior); and (3) the *judgment* is the doctor's decision after evaluating the case (to commit the person or not).

Research suggests that doctors are not very good at predicting whether harmful future behavior (such as suicide) will occur after they evaluate a person. Whether a person commits a harmful act soon after being examined by a doctor can depend as much on later unknown, unpredictable factors (such as whether a friend returns a telephone call) as on factors that can be measured during the examination.

Conclusion

Although involuntary commitment often is discussed in legal rather than ethical terms, it is important to remember that there are underlying ethical issues. Civil commitment involves confining—for days, weeks, or longer—an unwilling person who is not a convicted criminal. Such a seemingly unethical action requires clear ethical reasons to justify it. Whether or not the commitment is ethical may depend on reasons such as the degree to which the person's behavior is involuntary or irrational and on the process of predicting future behavior.

The American Psychiatric Association has developed a guideline for diagnostic evaluation, the *Diagnostic and Statistical Manual IV.* The manual emphasizes the importance of considering a number of factors—including symptoms, character or personality, medical status, drug and medication history, life events, and ability to function in society—before recommending a course of action that may include commitment to a mental hospital.

Related Literature

Ken Kesey's novel *One Flew Over the Cuckoo's Nest* (1962) portrays life in a mental institution, where at least one of the patients, McMurphy, is not mentally ill but is feigning illness in order to escape work on a prison farm. When he tries to challenge Nurse Ratched and her tyrannical use of medication and shock treatments to keep the patients under control, the results sometimes are hilarious and sometimes lead to disaster, such as the suicide of a patient. To finally control McMurphy, Nurse Ratched has him lobotomized. The book was made into a powerful movie (1975). Both the book and the movie challenge the power and appropriateness of psychiatry and mental institutions.

Anne Sexton's book of poetry *To Bedlam and Part Way Back* (1960) opens with a poem addressed to "You, Doctor Martin." She describes waiting in lines in the asylum while orderlies unlock doors and count patients as they go to dinner (where the silverware includes no knives in case patients are suicidal). She

spends her days making moccasins. In the poem *"Music Swims Back to Me,"* she recalls the tune someone was playing on the night she was committed to the mental institution. Another poem, "Noon Walk on the Asylum Lawn," portrays the narrator's paranoia. Even the grass blades seem threatening, and there is no safe place to hide from her enemies. The whole collection of poems gives voice to the mentally ill and describes their experiences "from the inside."

DRUGS TO TREAT MENTAL ILLNESS

psychiatric having to do with the treatment of mental illness

psychotropic affecting psychological function, behavior, or experience

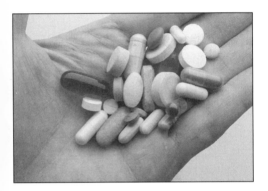

While drugs are successful in the treatment of some mental illness, their prescription and use have to be strictly monitored to protect patients' well-being.

Even without a diagnosis of mental illness some people still use drugs such as Prozac to give them a little boost. This off-label use of a drug is frowned on by doctors and manufacturers, but it is hard to control. People in their thirties and forties in the 1990s were referred to as "the Prozac generation."

symptom something that indicates disease or bodily disorder

Psychopharmacology is the study of drugs used to change people's mood, behavior, or mental functioning. While many drugs produce behavioral or psychological changes, psychopharmacologic agents are used specifically for such effects. These drugs have been available to doctors only since 1955.

Ethical issues arise when people overuse drugs, including prescribed dosages that are too large; when they use several drugs at once; and when they use drugs even though pharmacological treatment is not necessary. Additional ethical issues arise around whether **psychiatric** patients can consent to treatment and regarding the many negative effects from **psychotropic** drugs. The introduction of new drugs has greatly increased the cost of treatment, and the controversy over how drugs are sold raises further concerns. Psychiatric patients often lack the ability to protect their rights because of limited judgment or experience. Certain groups, including confused, demented, and retarded people—and children—are also likely to receive psychotropic drugs and are unable to protect themselves. This entry will also discuss drugs used to fight anxiety, as well as other drug classes. (Drugs will be referred to by their generic name with the American trade name in parentheses.)

Drugs are a multibillion-dollar business. The development of a drug in the United States or Canada may take ten years and hundreds of millions of dollars. Testing drugs on humans produces unknown risks, first to healthy volunteers and later to sick patients. It is not unusual for investigators conducting drug tests also to act as consultants to drug companies, possibly affecting their ability to be objective.

Classes of Mood-Altering Drugs

Antipsychotics. Antipsychotic drugs have made possible important changes in the treatment of the most serious psychiatric disorders, allowing the control of disabling **symptoms**, such as delusions (falsely held beliefs) and hallucinations (such as voices heard by the patient). Chlorpromazine (Thorazine) was the first antipsychotic introduced. Haloperidol (Haldol) is the most widely

psychosis a mental disorder with such symptoms as delusions or hallucinations that show a disturbed contact with reality such as schizophrenia or paranoia

schizophrenia a mental disorder characterized by a view of the world that does not match reality, sometimes causing odd behavior.

see also

Mental Health: Abuses of Psychiatry, Commitment to Mental Institutions

Woman with catatonic schizophrenia, a mental condition characterized by physical disturbances such as bizarre postures and lack of movement or expression.

receptor a molecule or chemical group in or on a cell that has an affinity for a particular chemical, molecule or virus

dopamine a form of the amino acid dopa that acts as an especially effective transmitter of impulses in the nervous system

neurological relating to the nervous system

used in the United States. Before antipsychotics came into use in 1952, treating **psychosis** was almost impossible. The usual treatment method was to hospitalize the patient—often indefinitely—which caused little or no change.

Schizophrenia. Schizophrenia is the illness that has benefited most from antipsychotic drugs. Schizophrenia affects 1 percent of the population worldwide, and most of these patients are treated with antipsychotics. These drugs are especially useful in treating hallucinations, delusions, and a variety of paranoid symptoms. They greatly reduce the patients' hostility and aggressive behavior, which can pose a great risk to the caregivers or patients' families.

Most schizophrenic patients first experience their illness in early adulthood, and relapses of psychotic behavior can occur for decades. Thus, after the first episode of psychosis is treated, long-term therapy is usually necessary. While a patient is experiencing a severe psychotic episode, there are few guidelines to determine a correct dosage of antipsychotics. Only a small percentage of schizophrenic patients improve without antipsychotic drugs.

Long-term effects. Despite their effectiveness, antipsychotics are far from perfect. They produce a number of uncomfortable—sometimes permanent—effects on the brain and behavior. Although schizophrenic patients are less psychotic and less dangerous with these drugs, they are still sick. They still experience withdrawal from society, loss of motivation, slowed thinking, and occasional bizarre behavior.

How antipsychotics work. Antipsychotics work by blocking dopamine receptors in nerve cells in areas of the brain that are known to determine psychotic behavior. These **receptors** receive the chemical **dopamine** from other nerve cells. Increased dopamine activity seems to make individuals more psychotic, while blocking dopamine reduces psychosis. In addition to this role in mental functioning, dopamine is also involved in muscular function and in the synthesis of norepinephrine, a chemical important in arousal and attention. Therefore, the blocking of dopamine produces uncomfortable muscular symptoms (stiffness, shaking, restlessness) and reduces alertness and attention. In addition, antipsychotics have a usually negative effect on receptors of many other brain systems.

Side effects. Antipsychotic drugs are not popular with patients because of their unpleasant side effects. Many patients who feel sedated, slowed, and uncomfortable stop taking their drugs. More important, a large percentage of patients who take antipsychotics over a long period of time develop tardive dyskinesia, a potentially permanent **neurological** condition that causes the mouth, face, neck, and body to move involuntarily. Each additional year of treatment with antipsychotics increases a patient's chance of developing tardive dyskinesia. After three years of therapy (common for schizophrenia), 15 percent of patients will develop this side

A depressed woman is huddled in a hallway corner. Depression is a widespread condition in the United States, affecting at least 6 percent of Americans at some point in their lives.

Mental Health: Electroconvulsive Therapy

The March 26, 1990, cover of *Newsweek* magazine depicted a Prozac capsule floating in the blue sky. The interest in these drugs comes in large part because many of them are very effective. The July 6, 1992, cover story of *Time* magazine was "Pills for the Mind."

psychiatrist a doctor who treats patients who suffer from mental illness

effect. Elderly, hospitalized psychiatric patients have the highest rates of this disorder, with many surveys finding over 50 percent affected.

Antidepressants. Antidepressants are widely used, reflecting the high rates of depression in the U.S. population. At least 6 percent of Americans experience depression at some point in their lives. The introduction of these drugs followed the discovery of antipsychotics and created a totally new approach to the treatment of depression. Their success has increased the understanding of depression as a medical disorder similar to diabetes or high blood pressure. Like antipsychotics, these drugs have reduced the need for hospital stays and shortened their length when they become necessary. Unlike antipsychotics, antidepressants work in many different ways; it is not understood at all how some of them work.

Newer uses for antidepressants. While norepinephrine was the main target of early antidepressants, newer drugs have greater effects on the chemical serotonin. Reduced levels of serotonin have been identified in suicidal behavior, obsessive-compulsive disorder, and some violent behaviors. Thus, as these antidepressants block the reuptake of serotonin, more serotonin becomes available to the receptors, relieving all of these symptoms.

Prescriptions multiply. Several drugs that increase the availability of serotonin specifically are available now: fluoxetine (Prozac), sertraline (Zoloft), paroxetine (Paxil), and clomipramine (Anafranil). These drugs have had a huge impact on the medical community and have been widely prescribed—seven million Americans take fluoxetine alone. All of these drugs are easy to prescribe and are used much more by primary care doctors than by **psychiatrists**. They are also expensive. Because depression can reoccur, many patients will require long-term therapy, perhaps for decades or more.

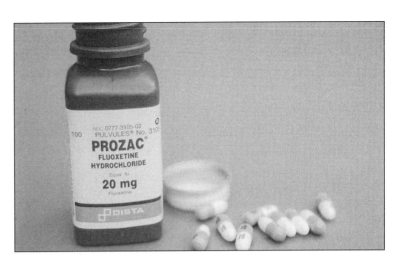

▶ Prozac is among the most successful antidepressants available.

Lithium is considered an antidepressant, but it is used much more widely to treat **manic-depressive disorder**. Although the way that it works is not fully understood, it can prevent manic depression from coming back, as well as treat severe episodes of both mania and depression.

manic-depressive disorder a mental illness characterized by alternating reckless and depressive states; also known as bipolar disorder

Antianxiety Drugs

Antianxiety drugs are used to treat primary **anxiety** disorders as well as secondary anxiety that results from a number of other medical conditions (such as a heart attack). Alcohol is the oldest antianxiety "drug."

anxiety intense and overwhelming fear or dread, often marked by physical symptoms (sweating, tension, increased pulse)

Early treatment of anxiety disorders. Barbiturates and **propanediols** were the first doctor-prescribed antianxiety drugs. These had sedative and anxiety-reducing effects, but they also slowed thinking and alertness. In the late 1960s, scientists developed the **benzodiazepines**, drugs that reduced anxiety but allowed patients to retain mental and physical functioning. These drugs include diazepam (Valium), lorazepam (Atvian), and alprazolam (Xanax).

barbiturate one of many derivatives of barbituric acid; used in making sedative and hypnotic drugs

propanediol compounds containing diols (OH-groups) and the organic substance C_3H_8; used in antianxiety medications

benzodiazepine one of many substances called aromatic lipophilic amines; used in making tranquilizers

Common but undertreated conditions. Anxiety and panic disorders, for which these drugs are mainly used, are the most common psychiatric disorders. Almost 15 percent of Americans experience anxiety disorders at some point in their lives. In one survey, 8.3 percent of the U.S. adult population reported being affected by an anxiety disorder in the previous six months. Because anxiety is so common, these drugs are widely used, even though fewer than one-quarter of the people suffering from anxiety ever seek treatment.

The flower known as St. John's Wort has been used to treat depression since at least the twelfth century. People who take St. John's Wort consider it very helpful in elevating mood. In Germany, St. John's Wort is so popular that synthetic antidepressants such as Prozac are very difficult to market. St. John's Wort can be purchased without a prescription at most American health food stores. Scientists claim that the effect of St. John's Wort has not yet been proven.

Ethical Dilemmas in Drug Therapy

A major ethical concern involves giving mood-altering drugs to patients who are so impaired by their condition that they are unable to make an informed choice about their treatment and its risks and benefits.

The Scream by Edvard Munch (1863–1944). This image has become an icon for panic and despair.

paranoid excessively fearful, often with delusions of being under attack

Consent to treatment. Antipsychotic drugs are prescribed mainly to psychotic patients, many of whom are **paranoid** and suspicious, especially about drugs they are asked to take. Doctors face a dilemma: Can patients reasonably accept or refuse treatment for an illness that prevents them from experiencing reality, causes them to suspect people who want to help, and may cause them to be a risk to themselves or others?

Refusal of treatment. All competent patients have a right to refuse treatment of any kind. Unfortunately, state laws have not clearly defined competency in patients who suffer from psychotic disorders. Competency hearings may delay decision making for weeks and are expensive for patients, doctors, and hospitals. If patients refuse to take medication and, as a result, become dangerous to themselves or others, state laws—and common sense—hold that medication can be temporarily forced if necessary. Forced medication involves giving drugs to patients against their will. In less extreme cases, patients participate in their treatment, but this cannot fully be considered free

A schizophrenic woman in a hospital psychiatric ward. The very nature of the disease makes it difficult for such a patient to understand and agree to the treatment recommended by doctors.

informed consent consent to medical treatment by a patient—or to participation in a medical experiment by a subject—after achieving an understanding of what is involved

Ethics and Law: Information Disclosure, Truth-Telling, and Informed Consent

separation anxiety in children, the distress felt because of disconnection from a parent or other important figure

will if this "choice" is made on a locked ward in a hospital. One study revealed that the most severely psychotic patients refuse treatment more frequently than less psychotic ones.

Continued treatment. As patients improve from severe psychotic episodes, the ethics of continuing drug therapy becomes less clear. Side effects of antipsychotics are unpleasant and may interfere with patients' work, social life, or sex life. The possibility of tardive dyskinesia, for example, is so great that patients should be specifically informed about it. Yet antipsychotics can prevent relapses that might require further hospitalization. Written permission may be obtained to show **informed consent**, but there is no substitute for a relationship in which the doctor supplies information in response to the patient's needs and wishes. Ideally, as patients respond to therapy, they become less psychotic and more able to participate with their doctors in making informed decisions about their treatment.

Drug therapy for children. Psychotropic drugs have clearly helped in treating such childhood and adolescent disorders as anxiety, refusal to go to school, and **separation anxiety**. For psychotic children, antipsychotics are often effective, though all of the side effects seen in adults can also occur in children. Informed consent in these situations generally involves informing the parents, who are responsible for authorizing treatment until the child reaches the age of eighteen.

ADDH. Attention deficit disorder with hyperactivity (ADDH) is a childhood condition of extreme muscular activity, concentration problems, restlessness, and impulsive behavior. Symptoms are generally worse at school, where the child may also become easily frustrated and aggressive, not get along with fellow students, and display poor academic performance. Antipsychotics and antidepressants have been used to relieve this disorder, but it is usually treated by psychostimulants, such as methylpenidate (Ritalin). The rate of success with methylpenidate is as high as 90 percent.

▶ A boy suffering from ADDH is being restrained by his mother.

Effect of drugs on children's performance. While ADDH patients' performance improves with stimulant drug treatment, so does the performance of normal children. There is some disagreement as to the role of these drugs in improving academic achievement. Some argue that the drugs improve teachers' ability to control the students in the classroom, not the students' academic skills. In addition, drug treatment makes no long-term improvement. Thus the overeager use of these drugs in classrooms may benefit the school more than the student.

From a treatment standpoint, drug therapy is continued as long as the child remains hyperactive, although many adults who had ADDH in childhood still benefit from taking stimulants. The drugs reduce appetite and disturb sleep. The decision to give a child these drugs is complicated, and the risks and benefits must be weighed for each child individually. Parents, teachers, and doctors should all have input in deciding when to start and stop treatment.

Summary

Mood-altering drug treatments are becoming more successful and are being more widely used. An understanding of these drugs and what they can do must go hand in hand with an awareness of their effects on patients and their individual needs. In most cases, appreciation of the ethical dilemmas surrounding drug therapy will lead the doctor to involve patients in decision making and to help patients and their families understand the risks and benefits of treatment.

Instead of taking the time to analyze the problems of patients through conversation and nondrug interventions, psychiatrists too often are reduced to prescribing pills. Moreover, the long-term adverse side effects of many of these drugs may not be fully explained to patients.

▶ An anxious-looking teenage girl sits apart from her group of peers. Chemical treatment of some psychological trouble has proved successful, but the ethical dilemma presented by mood-altering drug therapy remains.

ELECTROCONVULSIVE THERAPY

psychiatry a branch of medicine that deals with the treatment and prevention of mental illness

psychiatrist a doctor who treats patients who suffer from mental illness

Ernest Hemingway (1899–1961) was one of America's greatest novelists and short-story writers. He was awarded the Nobel Prize for literature in 1954. Hemingway, not unlike some other famous writers, struggled with some depression, but he could usually write his way through it. Unfortunately, Hemingway killed himself in 1961 after repeated series of high-voltage electroconvulsive therapy, which may have harmed his memory and left him unable to write.

Electroconvulsive therapy (ECT), commonly known as shock therapy, is a highly controversial and sometimes effective treatment in **psychiatry**. Some people believe that electroconvulsive therapy should be against the law because it can seriously harm a patient's memory. Others feel that ECT is a crude form of behavior control that **psychiatrists** often force patients to accept.

Years ago, doctors gave shock therapy in ways that are not used today. Higher amounts of electrical current and sometimes one or more treatments a day were given to patients. Without a doubt, this treatment harmed many patients.

The nature of the treatment itself has always seemed frightening to many people. Movies and novels have featured horrifying portrayals of shock therapy. The idea of passing an electrical current through the brain, causing a brain seizure and unconsciousness, is indeed very scary. That no one seems to know exactly how the treatment works adds to the fear. There are, however, many effective treatments in medicine whose ways of working are unknown. And there are probably many operations that would seem just as frightening as ECT if the ordinary person observed them. The basic question is, Can it be ethical to pass an electrical current through someone's brain, knowing that the person probably will suffer some memory loss?

electrode a conductor of electricity

schizophrenia a mental disorder characterized by a view of the world that does not match reality, sometimes causing odd behavior

Chemical shocking agents to treat mental illness have a long history. The first modern effort was described by L.J. Meduna in 1935. Meduna induced convulsions by intramuscular injections of camphor, and later by injections of Metrezol, a synthetic camphor. Insulin coma was another chemical shock treatment that followed soon after Metrezol. In 1950, the International Congress of Psychiatry accepted chemical and electric shock treatments as the best available for people with schizophrenia. Since the 1970s, it has been clear that schizophrenia is not helped by such treatments, and may even be worsened.

ECT Treatment

There are several excellent studies of the history of electroconvulsive therapy and its likely harms and benefits. The treatment basically consists of bringing about a brain seizure. This is done by attaching **electrodes** to the scalp and passing an electrical current through them for a fraction of a second. Treatments are usually given two to three times a week for two to four weeks.

Originally, doctors used ECT as a treatment for **schizophrenia**, but this use was eventually rejected as futile (of no benefit). Doctors now use ECT almost entirely with patients who are suffering from severe depression. Most psychiatrists use it only when drug treatment or psychotherapy (counseling) have not helped the patient.

Benefits of ECT. Most experts agree that ECT is effective in reversing severe cases of depression. Studies show recovery from depression in 75 to 85 percent of patients who receive ECT. In contrast, only 50 to 60 percent of depressed patients respond to antidepressant drugs. Patients who do not respond to drugs show the same response rate to ECT as do patients in general. ECT also works more quickly than drugs. With shock therapy, patients usually improve after about one week. Drugs, if they work, typically take three to four weeks, and sometimes longer. Most studies show that when ECT is applied to one or both sides of the head, it is equally effective. A few studies have found that the effects on average are not quite as good when ECT is limited to just one side of the brain.

Drawbacks of ECT. The main harmful effect of ECT is memory loss. ECT causes two kinds of memory loss. During the two to three weeks that doctors give treatments, memory and other thought functions are usually mildly to moderately harmed because of the ongoing seizures. Also, years after the treatments, patients are unable to recall many events that took place shortly before, during, and after the two- to three-week course of treatment.

The more important and controversial question is, How often does ECT cause a *permanent* loss of memory? If and when it does, it is possible that the treatment has damaged parts of the brain that control memory function. The effects of ECT on memory have proved to be a difficult research problem, despite dozens of studies that scientists have carried out. Among the many difficulties involved is the fact that depression itself often causes memory loss.

Ethical Issues

Is shock therapy so harmful that it should be against the law? Few experts take this position. ECT has an extremely small risk of causing death. It also has a risk of causing ongoing minor memory loss and a small risk of causing ongoing serious memory loss. ECT is frequently used when it is the only available treatment. Often, doctors use it when the patient is suffering terribly and may be at risk of dying.

Ethics and Law: Information Disclosure, Truth-Telling, and Informed Consent

Many critics of ECT believe that it should be used only as a last resort, when drugs and other forms of therapy have proved ineffective, and only if the patient feels that his or her depression is severe. Legal philosopher Gerald Dworkin defends these treatment guidelines:

1. Methods that support self-respect and dignity are to be encouraged.

2. Methods that are destructive of the ability of individuals to reflect rationally should not be used.

3. Methods that affect the personal identity of individuals should not be used.

4. Methods that rely on deception are to be avoided.

5. Methods that are physically nonintrusive are to be preferred over intrusive ones (such as ECT).

6. Methods that work through the active participation of the mental and emotional capacities of the patient are preferred over methods that shortcut or blunt these capacities.

As with all other medical treatments, doctors should explain the possible harms and benefits of ECT to the patient during the consent process. Doctors need to mention the risk of death and of chronic memory problems. In turn, the patient needs to understand and weigh the information. The patient also needs to ask the doctor questions and to be allowed enough time before making a decision.

Doctors often suggest shock therapy to patients only after other treatments have failed. But ECT has several advantages over other treatments: (1) it works more quickly; (2) it works in a higher percentage of cases; and (3) it does not have the dangerous side effects of many antidepressants.

Many people believe that if a patient is given enough information about the treatment and understands and appreciates this information, then the patient's decision to have—or not to have—electroconvulsive therapy is valid. As long as the patient's choice is not forced, the decision should be respected.

Related Literature

Although Janet Frame's *Faces in the Water* (1961) is a novel, there is no doubt that the narrator, Istina, knows what she's talking about firsthand. She describes the disorientation and memory loss caused by her many shock treatments. The narrator discovers herself finally in reading and writing. She can recover some of her identity by keeping good company with the books she chooses; she can also create her own poetry.

MENTAL HEALTH AND MENTAL ILLNESS

This entry consists of two articles explaining various aspects of this topic:

THE MEANING OF MENTAL HEALTH Ideas of "health" and "mental health" depend on cultural, social, political, and economic concepts. What is "normal" varies greatly from one historical period or culture to another. The histories of health and mental health are many-sided and far-reaching. Traces of their past are still seen in problems and controversies that focus on how mental health and mental illness are defined.

Historical and Philosophical Background

The seventeenth century saw the continuing spread of individualism and a reliance on reason. Irrationality—a break with reason—became the key standard for madness. According to the twentieth-century French philosopher Michel Foucault, the seventeenth-century insane asylums were filled not with "mad" people but with individuals who had violated reason and socially acceptable behavior. Foucault made people aware of the possible links between social, cultural, political, and economic special interests, on the one hand, and **psychiatric** institutions, ideas, and practices, on the other. He also alerted people to the problems in defining mental health and mental illness.

The period from 1600 to 1750 was a turning point in many ways. Its scientific models, reliance on reason, and social, cultural, and economic developments paved the way for the nonreligious outlook and system of capitalism that followed in the West. The coming age would require and give rise to new notions and ways of well-being, dysfunction, and distress. It would also produce a new medical specialty: **psychiatry**

Development of psychiatry. If the seventeenth century spawned new ideas about "wellness" and abnormal functioning, then the eighteenth century gave rise to modern ideas about mental health. In early and mid-eighteenth-century Great Britain, a new breed of doctors began devoting their practices to madness. The most brilliant of these "mad-doctors," Alexander Crichton, influenced Philippe Pinel, who has been called the father of psychiatry. Pinel prospered after the French Revolution of 1789 and in the early nineteenth century. Until then, madness had not been generally treated in medical institutions. Asylums typically fell under the management of people outside the medical profession, with doctors serving only as consultants. Pinel's training was both psychological and medical. This joint orientation led him to understand and classify mental illness in a new way.

psychiatric having to do with the treatment of mental illness

psychiatry a branch of medicine that deals with the treatment and prevention of mental illness

"I proceed, gentlemen, briefly to call your attention to the present state of insane persons within the Commonwealth, in cages, closets, cellars, stalls, pens—chained naked, beaten with rods, and lashed into obedience."

—Dorothea L. Dix (1802–1887), mental health reformer and superintendent of women nurses for the Union Army, in a memorandum to the Massachusetts legislature (1841).

▶ A restraint device at the New York Insane Asylum during the late nineteenth century, with a patient confined in its enclosed space.

In the eighteenth century, people in Philadelphia paid an admission fee to gawk at insane patients chained to walls in mental asylums. Insanity was considered the result of demon possession, and the insane, therefore, as evil. The American Quakers (Society of Friends) took a different view. By the 1800s, they succeeded in establishing "moral treatment" for such patients, who were allowed to walk around freely within the institution and to work in gardens. The Quakers revolutionized care of the insane by treating them with respect, dignity, and hope.

psychiatrist a doctor who treats patients who suffer from mental illness

As the twentieth century approached, the number and size of asylums grew tremendously. Treatment became custodial, marked by caring for inmates and protecting them rather than by seeking a cure. Psychiatry's interest in classifying diagnoses and in the results of autopsies contributed, in Foucault's view, to the handling of the patient as an object. The rise of custodial psychiatry reflected many social changes in the United States. These changes included a rapidly increasing population; greater social and geographic mobility; replacement of small communities by urban centers overflowing with immigrants; the continued weakening of organized religion; a movement toward capitalism and consumerism; increased individualism and a lessening of local charity; and society's generally changing moral attitudes, customs, and beliefs. Together, such factors made psychiatric therapy unworkable and led to further changes in popular ideas about the individual in wellness and illness. Communities and even families transferred responsibilities for their mentally disturbed members to large central institutions.

Such institutions came to house many people who were merely elderly, socially different but not criminal, and economically unproductive. Certain "diagnoses" of the time—such as "old maid," "vagabond," and "eccentric character"—would be laughable if they had not also been enforced by society. State hospitals usually fell under the guidance of social agencies that dealt with individuals who were social and economic outsiders.

Drawing on such historical sources, as well as on modern events, a group of social scientists and political philosophers has criticized the self-serving features of psychiatry, including its notions of mental health and mental illness. These notions incorporate biases based on gender, class, and ethnic background. These critics further suggest that psychiatry and **psychiatrists** tend to have an investment in maintaining the existing system of mental health and mental illness.

Twentieth Century

By 1910, events relating to mental health and mental illness were gathering steam. The Mental Hygiene Movement—a joint venture of psychiatrists and nonmedical workers—had been formed in Boston in 1909. Though the movement was first begun in order to improve the plight of the severely mentally ill (formerly the "mad"), its concerns shifted quickly toward mild-to-moderate psychiatric problems and to community mental hygiene. This shift, in turn, led to the growing community mental-health movement of the 1950s, 1960s, and 1970s. This movement, together with the new therapies, fueled public preoccupation with mental health.

Meanwhile, psychiatrists in the United States had started to move toward acute-treatment psychiatric facilities and wards in general hospitals that treated not the chronically ill but rather patients suffering from brief but severe crises, neurotic symptoms, and

outpatient a patient who is treated at a hospital but does not stay there overnight

Thomas Szasz wrote: "In actual contemporary social usage, the finding of a mental illness is made by establishing a deviance in behavior from certain psychosocial, ethical, or legal norms." Szasz believes that mental illness is a label used by societies to designate nonconformists as sick. Partly to defend against such critics, modern American psychiatry emphasizes objectively measurable differences in the brain chemistry and structure of individuals with severe mental illness. This organic approach allows psychiatry greater acceptance as a science. But critics of the organic approach complain that too many psychiatrists do nothing more than prescribe new drugs, ignoring other therapeutic methods such as psychoanalysis.

atonement making up for wrongdoing

personality problems. Work with **outpatients** continued to grow as well. Clinical psychology and social work became professions. General medicine's public-health and preventive wings expanded. These developments led many critics to speak of medical and psychiatric "imperialism," the "medicalization" of society, and similar descriptions with negative judgments. Indeed, it was true that some physicians had recommended that doctors and medicine replace priests and religion as society's moral judges.

Definitions of "health" as broad as the World Health Organization's (1991) "state of complete physical, mental, and social well-being" and classifications of "mental disorder" as extensive as those of the American Psychiatric Association (1987, 1994) seem to give substance to some of these accusations. Even aspects of once "normal" aging are now regarded and treated as diseases. Doctors and the public alike view death itself as all but a potentially preventable disease.

Conclusion

Given the historical and cross-cultural differences in how to classify someone as mentally ill, it makes little sense to seek timeless and universal definitions of health, illness, or even disease. All that can be said is that however we understand mental health and mental illness, these ideas point toward forms of distress, disability, and well-being that are real and human concerns.

CONCEPTIONS OF MENTAL ILLNESS In discussing ideas of mental health and mental illness, we might ask ourselves the following questions: What counts as madness? acting out? batty? bizarre? breaking down? cracked? crazy? daft? demented? depressed? deranged? erratic? frenzied? gaga? hysterical? idiotic? inane? insane? irrational? imbecile? jerky? kooky? lunatic? lulu? manic? melancholic? mentally ill? moronic? neurotic? nuts? off one's rocker? paranoid? possessed? psycho? psychotic? raving? schizoid? touched? unglued? wacky? weird? yo-yo? And why are there so many terms for being mentally ill? In his play *Hamlet*, Shakespeare suggests that the idea of mental illness cannot be defined: "to define true madness, What is't but to be nothing else but mad?" (2.2.93–94). Yet Western law, medicine, philosophy, and sociology have tried to define this elusive and puzzling thing called madness.

In traditional societies, illness tends to be thought of in ethical and religious terms. In other words, illness is considered a punishment for a moral wrongdoing or a sin. A "diagnosis" becomes a determination of the ill person's violation. "Treatment" involves **atonement** or prayer. Ethical and religious accounts of illnesses are common in the Hebrew Bible. In the New Testament, however, Jesus answers the question, "Rabbi, who hath sinned, this man or his

parents, that he should be born blind?'' by rejecting the ethical and religious association of illness with wrongdoing. Jesus says: ''Neither hath this man sinned, nor his parents'' (John 9:1–3). In general, Christian societies have followed Jesus' words and accepted the idea that most disabilities and illnesses have causes unrelated to ethics or religion, but madness has tended to be an exception. Ethical and religious accounts of madness continued into the sixteenth and seventeenth centuries. By the eighteenth century, doctors and most educated people treated madness as a form of illness.

Madness as Illness: The Hippocratic Model

The Western tradition of treating madness as an illness has its roots in the collection of texts from the fifth, fourth, and third centuries B.C.E. written by doctors associated with the school of Hippocrates. According to Hippocrates' followers:

1. Madness is a form of illness. Therefore, (a) it is a natural phenomenon, not a punishment for an ethical or religious transgression inflicted by demons, gods, or spirits; (b) no special shame is attached to madness—it is not a fall from grace but a physical state like ''pain, grief and tears'' and arises from ''the brain alone''; (c) it is involuntary—people who are mad cannot control their actions; (d) those afflicted are sick and thus excused from normal duties; and (e) it is to be diagnosed and treated by doctors, not priests or tribal healers.

► This manuscript illustration shows Hippocrates teaching his students.

Homosexuality was considered a mental illness by American psychiatry until this view was reversed in the 1970s due to political and moral pressure from the gay community. Social prejudices against women have resulted in a tendency in American psychiatry to label some female behaviors as illnesses.

Enlightenment a movement in eighteenth-century Europe that used reason as its guiding rule

In the 1970s, religious conversion to unpopular new religious movements was sometimes labeled as mental illness. This resulted in serious violations of the First Amendment of the United States Constitution, which guarantees freedom of religion. One of the violations of this freedom occurred in the context of forced deprogramming, in which it was falsely assumed that adult converts to new religious movements were mentally ill. Throughout the course of history, converts to new and socially unaccepted religions, from Quakers and Mennonites to Mormons and Hare Krishnas, were designated as mentally ill by some professionals, often pressured by parents resentful of a controversial new direction in an adult child's life. The topic of religion and madness is discussed in Roy Porter's *A Social History of Madness* (1989).

2. The mad are patients. They enjoy the special protections provided for patients within the Hippocratic texts, and in all writings on Western medical ethics that have followed.

3. Madness is a symptom of a disease internal to the organism. The followers of Hippocrates treated mad behavior as a symptom of an underlying imbalance of the four basic humors—wetness, dryness, heat, cold—within the brain.

4. Madness, like other diseases, can be diagnosed in terms of its causes. In the Hippocratic writings there are several attempts to classify mental illnesses based on their patterns of development, their cause, or their abnormal traits.

Antipsychiatry's Critique of Mental Illness

"Antipsychiatry" is a term used to describe the views of some historians, philosophers, psychiatrists, and sociologists. These doctors and scholars challenge one or more aspects of the medical model of madness as set forth by psychiatrists.

The European school. Michel Foucault "reveals" that psychiatry's outward caring for humanity only masks the **Enlightenment**'s repressive intolerance of unreason. Enlightenment thinkers, he writes, considered unreason to be "unchained animality [that] could be mastered only by discipline and brutalizing."

The British psychiatrist R. D. Laing developed an entirely different type of antipsychiatric critique. In his book *The Politics of Experience* (1967), Laing argues that schizophrenia cannot be an illness because it does not interfere with normal functioning. On the contrary, Laing believes, schizophrenia is really a coping strategy that helps people survive unlivable situations. It is a way by which the mind restructures itself. Schizophrenia can serve as a successful form of self-therapy. According to Laing, it is ironic that traditional attempts by psychiatrists to treat schizophrenia can actually disrupt the patient's own attempt at a cure.

The American approach. Thomas Szasz is the best-known American writer on antipsychiatry. He argues that psychiatrists who are imposing the sick role on patients are not practicing scientific medicine. Medicine is scientific because its diagnoses rest on illnesses that can be observed by a doctor. Thus a diagnosis of a blood clot or ulcer can be confirmed by X ray or autopsy. Physical illnesses, like diabetes, can be diagnosed by tests even when there are no symptoms of the disease. But there are no similar tests for mental illness. Other people's minds can never be explored objectively. No one can examine anyone else's thought processes by

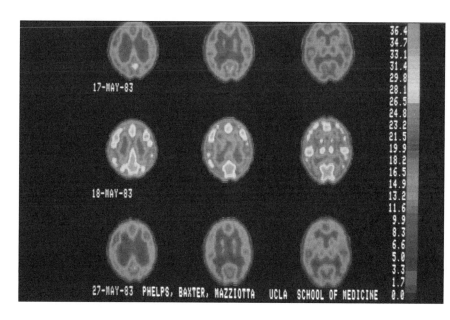

 Bipolar disorder shown in a series of positron emission tomography (PET) scans.

psychologist a nonmedical professional who treats patients suffering from emotional problems

Mental Health: Abuses of Psychiatry

Mental Health: Commitment to Mental Institutions

X ray or autopsy. There is no way of demonstrating that a psychiatric diagnosis is wrong. Scientific explanation, however, is always capable of being corrected, improved, or reformed. So, Szasz concludes, psychiatry is not a branch of scientific medicine.

Impact of the new theories. As a movement, antipsychiatry had an enormous influence in the 1960s and 1970s. These decades were concerned with civil rights and empowering oppressed peoples. Reformers seeking to have the mentally ill released from asylums and hospitals and treated in community programs made use of the antipsychiatrists' ideas. Their theories were also used by **psychologists**, social workers, and others who wanted to topple a mental-health power structure ruled by psychiatrists. At the same time, mounting evidence seemed to support the antipsychiatrists' claims. Diaries from Eastern Europe showed that Soviet psychiatrists were imprisoning in mental institutions citizens who dared to speak out against the regime. These Soviet psychiatrists seemed to be confirming the antipsychiatrists' charge that the term "mental illness" was being used to justify the imprisonment of people who were socially undesirable, not mentally ill.

Problems resulting from reforming the system. Even as historians were rethinking the antipsychiatrists' picture of psychiatry's past, problems were beginning to appear. Antipsychiatrists rejected the insanity defense in law; instead, they joined with civil libertarians to urge tightening the rules for involuntary commitment to mental institutions. They argued that a diagnosis of mental illness should not be sufficient grounds for committing someone involuntarily. They felt that individuals considered to be mentally ill should be committed to a hospital against their wishes only if the law determined that they posed a danger to themselves or to others.

These reforms were adopted and, throughout the Western world, the walls of the asylums came tumbling down. Thousands of

During World War II, sodium pentothal was widely used for its hypnotic effects, helping repressed people to recover and communicate lost memories. Contrary to various espionage films, sodium pentothal does not compel people to speak about something they wish to hide. It is not a "truth serum," and no such drugs exist.

mentally ill people who had been committed without their consent were found to have been illegally institutionalized. They were discharged into their communities. The results were not as positive as the antipsychiatrists had hoped. Once patients left the hospital, they were free. They were free of medical supervision, free not to report to clinics for therapy or counseling, free not to take their medicine. These patients tended to become ill again and to be problems for their communities. Unable to cope with the outside world, these former mental patients often became homeless and easy targets for criminals. Ironically, one of the principal effects of letting these mental patients go free was to give new life to the idea of the asylum as a safe place—a quiet haven from the world for mad people.

Related Literature

Peter Shaffer's play *Equus* (1974) takes place in a mental hospital where a psychiatrist is trying to help a mentally ill seventeen-year-old stableboy who has blinded several horses with a hoof pick. As the doctor works to understand what circumstances led to the teenager's brutal action against the horses, the "therapy" gradually reveals the complex causes, including the teenager's confusion of Christianity and his own worship of his horse god, Equus. The psychiatrist finds himself envying the passion of his sick patient and wishing that he had such powerful feelings himself.

Shakespeare gives us two vivid portraits of mentally ill people: Lady Macbeth from *Macbeth* (1606) and Lear from *King Lear* (1605). After pressuring her husband to kill Duncan, and helping him with the murder, Lady Macbeth cannot get rid of her guilty feelings. She walks in her sleep, madly trying to wash off bloodstains that she imagines are still on her hands. Macbeth asks the doctor if he is able to "minister to a mind diseased," but the doctor cannot help her. Lear may be suffering from some dementia related to his old age, but the cruel actions of two of his daughters who refuse him lodging certainly compound his problems and drive him to raging fury as a storm sweeps across the heath. Lear's own internal storms match the weather.

Population

EPIDEMICS

infectious spread by a virus or other microorganism

noninfectious pertaining to illnesses that cannot be spread from one person to the next

plague a highly contagious, often fatal, disease transmitted by fleas that have been infected by rodents; also called bubonic plague

endemic found in a certain region or among certain parts of a population on a regular basis

The eighteenth-century novelist Daniel Defoe wrote *A Journal of the Plague Year*, in which he reconstructs the Great London Plague of 1665. He includes numerous references to the services of doctors who, by trying to help others, became themselves victims of the plague. Also in the 1600s, doctors in Florence and other Italian cities entered plague-infested areas to help the sick, and many doctors died.

symptom something that indicates disease or bodily disorder

Epidemics are concentrated outbursts of diseases that are infectious or **noninfectious**. Epidemics cause a huge number of deaths, and they affect large numbers of people living in a particular space during fairly short periods of time. Epidemics probably first appeared among human beings during the "Neolithic Revolution," roughly eight to ten thousand years ago. This was a time when people began to domesticate animals, practice farming, and settle into towns and villages, causing a great many people to live in one area.

Ancient and Medieval Times

Texts attributed to Hippocrates (460?–377? B.C.E.), the famous Greek physician, indicate the presence of such diseases as tuberculosis (an infectious disease, usually of the lungs), malaria (a disease spread through the bite of an infected mosquito), and influenza (the "flu") in the population of ancient Greece. The Greek historian Thucydides (460?–400? B.C.E.) gives the first full description of the **plague** in Athens (430–429 B.C.E.) in his history of the Peloponnesian War. The increase in trade brought about by the growth of the Roman Empire, which flourished from 27 B.C.E. until 476 C.E., caused diseases to spread to many places. There were huge epidemics in the Mediterranean area (165–180 C.E. and 211–266 C.E.).

In Europe and Asia, diseases such as smallpox (a highly contagious virus causing pockmarks on the skin) and measles gradually became **endemic**. But there were still occasional epidemic outbursts. Periodic epidemics of plague continued. The most serious epidemics occurred in the fourteenth century, when perhaps as much as one-third of the population of Europe died.

Responses to Epidemics Before the Nineteenth Century

The ancient Greeks and Romans, for the most part, believed that epidemics entered communities from outside. Thucydides described the plague that struck Athens as having arrived by sea. This belief supported the way in which people viewed epidemics in medieval Europe. In some epidemics, such as the Great Plague of London in 1665, victims were kept in their own houses. These houses were then marked with a red cross to warn healthy people not to enter. Diseases were recognized as such only after **symptoms** began to appear. The idea of the carrier who has no symptoms but who "carries" a disease to others was unknown.

 The Plague of Florence

People's reactions to epidemics included fleeing infected areas and disobeying public-health orders, as well as attacking minorities and the poor. As plague spread through Europe in 1348–1349, rumors that the Jews were poisoning water supplies led to widespread **pogroms**. Over nine hundred Jews were killed in Erfurt, Germany, alone. Such actions reflected a general feeling, reinforced by the church, that plagues were visited upon human beings by an angry God who was disgusted by immoral behavior, the lack of religion, and the tolerance that people were showing toward nonbelievers.

State, popular, religious, and medical responses to epidemics remained fairly constant well into the nineteenth century. The medical understanding of plague continued to rely heavily on theories surrounding the humors. Humors were thought to be the four fluids in the body responsible for a person's health and temperament. They were blood, phlegm, choler (yellow bile), and melancholy (black bile). So great was this belief in humors as the agents of epidemic diseases that doctors centered their treatments on bloodletting and other therapies designed to restore the balance of the humors in the patient's body.

Help from the state played a role in reducing the impact of smallpox, the other major killer disease of the period after plague. The spread of smallpox was first reduced by **inoculation**. Mandatory programs of cowpox vaccination (cowpox is a virus in cows from which smallpox vaccine is made) brought about a marked reduction in the impact of smallpox on nineteenth-century Europe. Despite problems with some of these new control methods, which sometimes included accidentally spreading the disease, vaccination programs were the first major achievement of "medical policing."

pogrom an organized massacre of a minority group, especially Jews

Edward Jenner (1749–1823) was an English doctor who in 1796 developed a vaccination for smallpox. When this method was introduced to North America, President Thomas Jefferson helped to popularize it by inoculating his own family and about two hundred others.

inoculation the injection of a mild form of a disease into a person to build up immunity to the disease

The Impact of Cholera

Such theories and practices were brought into question by the arrival in Europe and North America of Asiatic cholera, an infectious disease of the intestines. Reaching Europe by the end of the 1820s, cholera was spread further through unsanitary and overcrowded living conditions in the rapidly growing towns and cities of the new industrial era. Cholera primarily affected the poorest in society as a result of these living conditions, as well as of unclean water sources and poor methods of waste disposal.

Cholera epidemics affected the United States in 1832, 1849, and 1866. (Each time, the disease came from Europe after a major political conflict—brought over by fleeing immigrants.) State, popular, and medical responses to the disease in 1830–1832 were unchanged from earlier reactions to epidemics. Authorities imposed **quarantine** regulations, military troops took up stations around the infected areas, victims were isolated, and hospitals prepared. But many, including the increasingly powerful industrial and trading interests, opposed such strict measures. Liberals felt that medical policing involved too much interference with individuals' freedom. Eventually, these people forced the state to retreat from fighting cholera by the time of the next epidemic, in the late 1840s. In addition, medical theories about **contagion** were brought into question by the failure of quarantine and isolation to stop its spread in Europe.

It is important to point out that until the 1880s, many doctors thought that cholera was caused by a "miasma"—a vapor rising from the ground under certain weather conditions. Doctors believed that cleaning up the cities could prevent the vapor by keeping the source of infection from getting into the soil. This idea spawned sanitation reform in Europe and the United States during this period.

quarantine keeping people who have been exposed to a contagious disease away from others; a state of forced isolation, often done to prevent the spread of a disease

contagion the spreading of disease by contact

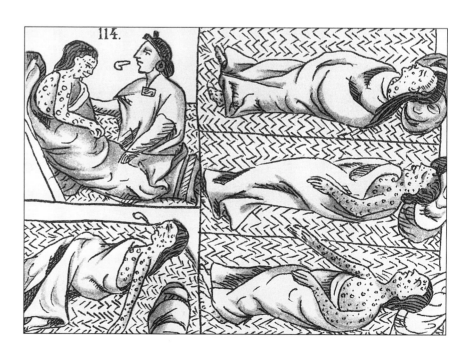

▶ A sixteenth-century drawing in the Codex Florentino of Aztec smallpox victims.

The rise of the medical profession, with its training and code of ethics, ensured that doctors were more active in treating victims of nineteenth-century epidemics.

The "Bacteria" Revolution

Cholera was only the most dramatic example of numerous infectious diseases that spread as a result of poor hygiene, city growth, overcrowded living conditions, and improved communications in the nineteenth century. Typhus (an infectious disease causing red spots on the skin), typhoid (a disease spread by infected food or water), diphtheria (a contagious disease that can block breathing), yellow fever (a tropical disease carried to people through the bite of an infected mosquito), tuberculosis, malaria, and syphilis (a sexually transmitted disease) continued to have a major impact. Even smallpox returned on a large scale during the Franco-Prussian War of 1870–1871.

Treatment for these diseases continued to fail. But the rapid development of microscope technology in the last quarter of the nineteenth century allowed medical science to discover the agents of many infectious diseases in humans and animals: bacteria. These discoveries ended the mistaken belief in "miasmas" as causes of epidemics. Instead, they generated the modern theory of the contagious nature of disease.

The "bacteria" revolution produced an age of increased state control over the spread of disease. Many countries introduced laws making the reporting of infectious diseases mandatory. Hospital building programs in the second half of the nineteenth century simplified the isolation of contagious patients in hygienic conditions and so prevented further spread of the disease.

The Decline of Epidemics

The provision of clean water supplies and effective sewage systems reflected growing community pride and the middle-class desire for cleanliness. It made epidemics such as the outbreak of cholera that killed more than eight thousand people in Hamburg, Germany, in little over six weeks in 1892 increasingly rare. Just as important were improvements in personal hygiene, which indicated general social trends as well as the growth of modern medicine in western Europe and the United States.

Such developments reinforced the idea that oppressed minorities and the poor were carriers of infection. These people were now blamed for ignoring official calls to maintain high cleanliness standards, even though their living conditions frequently made this difficult. In Europe and North America in the second half of the twentieth century, epidemic infectious disease was regarded as indicating an uncivilized state of mind. It was also considered characteristic of nonwhite populations in parts of the world outside the West.

Louis Pasteur (1822–1895) developed methods for destroying bacteria in milk (pasteurization) and for preventing rabies.

Robert Koch (1843–1910) identified the bacteria that cause tuberculosis (1882) and cholera (1884).

Epidemics of the Late Twentieth Century

In the 1980s, the discovery of a new epidemic, known as acquired immunodeficiency syndrome (AIDS), once again raised the ethical problems faced by society and the medical profession in the past. Lack of medical knowledge of the syndrome itself and of infection risk from contact with blood or other body fluids raised the question of whether doctors had a duty to treat AIDS sufferers even though there was no cure. The evidence of the overwhelming majority of past epidemics for which there was also no known cure seemed to be that medical treatment, even in the Middle Ages, could help suffering under some circumstances. Such medical treatment was therefore a duty of the doctor.

If some of the ethical problems surrounding the AIDS epidemic were fairly new, the question of mandatory public-health measures was a very old one. Like the victims in many previous epidemics, AIDS sufferers tended to come from already stigmatized groups: gays, drug abusers, prostitutes, Haitians, and Africans. On the one hand, the ability to screen these high-risk groups for the presence of the virus, HIV, that causes the disease, raised the possibility of required screening, quarantine, and isolation. On the other hand, individuals publicly identified as HIV-positive generally found it difficult or impossible to stay employed, to get life or health insurance, or to avoid being evicted from their homes. It was wrongly assumed that AIDS would not affect the heterosexual, **monogamous**, drug-free majority of voters.

> In 1986, the American Medical Association (AMA) published its *Statement on AIDS*, which announced that not every doctor is emotionally capable of caring for AIDS patients, and is free to make alternative arrangements by referring the AIDS patient to another doctor. This weak affirmation of a duty to treat led to intense criticism of the AMA. In 1988, the AMA, through its Council on Ethical and Legal Affairs, stated that "A physician may not ethically refuse to treat a patient whose condition is within the physician's current realm of competence solely because the patient is seropositive." By 1991, 80% of American doctors acknowledged a duty to treat HIV-infected patients.

Fortunately, new combinations of antiviral drugs are encouraging hope for an eventual AIDS cure, but the future remains bleak for AIDS sufferers around the world.

Conclusion

The history of epidemics suggests that society's responses have usually included using certain groups as scapegoats. From the Jews

see also

General Topics: AIDS

monogamous having only one sexual partner during a period of time

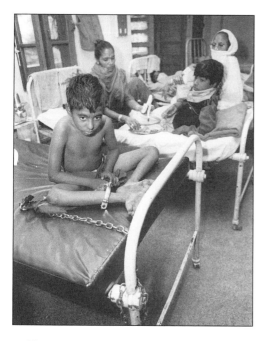

A young Indian boy is chained to a bed in a New Delhi hospital. The child shows symptoms of the plague. By Indian law, any person with such symptoms can be taken into custody in an effort to stop the spread of the highly infectious disease.

> "Typhoid, cholera, yellow fever, summer diarrhea, dysentery, tuberculosis, anthrax, intestinal worms, and many other diseases are still with us and could again return as plagues of old. During the garbage strike in New York in 1968, public health officials geared up to give massive immunizations against typhoid fever, so great was the potential hazard from garbage in the streets."
>
> —William E. Small, editor, *Biomedical News*, 1974.

massacred during the Black Death (plague) in medieval Europe to the minorities whose lifestyles were widely regarded as responsible for the spread of AIDS in the 1980s and 1990s, such groups have been subjected to exclusion and hostility during epidemics. Often, these groups have been the very people who have suffered most from the disease they were accused of spreading.

Clearly, any long-term solution to epidemics must be more than medical. Public-health measures are also highly political. Most politicians invest resources in educating the public and in taking other preventive measures only when they believe that the majority population is at risk. This means that these measures are put into effect only when public leaders consider the wider social and cultural context within which a disease operates.

Related Literature

Edgar Allan Poe's story "The Masque of the Red Death" (1838) tells of a group of rich people who try to escape the epidemic by taking an extended vacation in the castle of Prince Prospero. To entertain themselves while they are in this apparently safe retreat, they plan a costume ball. As the guests are all dancing and partying, a stranger appears dressed in a costume covered with red blood. When the host confronts the stranger, the host falls dead; when the revelers attack the intruder, they find there is no body within the costume, and they, too, start dying. Poe's symbolic tale warns that epidemics cannot always be escaped, even by the rich nobility.

In Chaucer's famous *Canterbury Tales* (1390), the characters tell each other stories as they travel on their pilgrimage to Canterbury Cathedral. One of the storytellers, the Pardoner, makes money by selling fake relics and indulgences to sinners who want to pay for their forgiveness. He tells a tale of three men, drunk rioters, who pledge to each other they will go out to "slay this false traitor Death," because he "hath a thousand slain in this pestilence." The plague is spreading throughout the country, and the three drunk men boast they can kill Death himself, and thus save everyone. Death, in the guise of an old man, sends the three men to a tree under which they find a pile of gold coins. It does not take long for their greed to provoke the men into killing each other, as each one tries to claim all the gold for himself. The tale reveals the spiritual or moral death in the three men as well as in the teller of the tale, the Pardoner.

FOOD POLICY

malnutrition an unhealthy condition of the body caused by not getting enough food or enough of the right foods

famine a widespread lack of food that causes many to starve

per capita for each person

It takes three pounds of grain to produce a pound of chicken, five pounds of grain to produce a pound of pork, and ten pounds of grain to produce a pound of beef. On average, every American indirectly consumes a ton of grain each year. If Americans ate less meat, the grain saved could be used to feed millions of people in other countries.

The world has always known hunger and **malnutrition**. But only recently has the world discovered that **famine** and starvation are avoidable. The technical ability to feed humankind exists. The main ethical concern focuses on the hunger from which more than 10 percent of the human race suffers constantly. The field of ethics also concerns itself with specific hunger events, such as famine, that affect certain populations at particular times.

The Condition of World Hunger

Most experts agree that the world produces enough food to feed every person on Earth. This production is expected to continue. Yet nearly 800 million people are undernourished and have to struggle every day to survive.

Past efforts to improve the international food system have failed. These efforts centered on increasing food production rather than on addressing the social, political, and economic forces that keep people hungry. If every person on Earth received the two pounds of grain **per capita** that the world produces every day, everyone would have at least a satisfactory diet. Increasing food production is an important goal, but the distribution of that food is at least as important. Because peasant producers and farm laborers are most at risk of hunger, increasing their food production and earnings is important. The key is to elevate the people who live in poor, rural areas of the world to living standards that are above basic survival conditions.

The 1974 United Nations World Food Conference met to deal with this problem. The conference concluded, "Every man, woman, and child has the inalienable right to be free from hunger and malnutrition in order to develop fully and maintain physical and mental faculties."

▶ Refugees waiting for food at a distribution point in Somalia.

Malnutrition is very harmful. Brain size is reduced and can result in permanent deficiency of intelligence. Physical vigor and size are lowered permanently. Nutritional deficiencies can cause specific diseases. For example, lack of Vitamin A causes blindness and lack of iron causes anemia.

Despite some progress in a few developing countries, the world continues to face the same ongoing and dangerous food crisis that has existed for many decades. While some countries struggle to pay more attention to food policy and to increase investment in agriculture, in many cases their populations continue to grow faster than their food production.

United Nations and World Bank (1986) data suggest that food production will continue to meet the demand for food, even if the world's population continues to grow. But the numbers do not reflect the different amounts of food consumed in each country or the ongoing hunger that comes from unequal incomes between and within countries. Most people in the industrial world, as well as some better-off minorities living in developing countries, consume nearly four times as much grain per capita as do poor people in developing countries. Poor people in poor countries eat their grain directly (for example, as bowls of rice). They get less food less often, and the food they do eat is less nutritious. Their right to food is violated. The violation of the right to have food, then, is the central problem of the world's food policy. Why is this right not guaranteed?

The Supply Side

Limited land. Roughly one acre of land is farmed for every person on Earth. There is probably an equal amount of land still available for farming. But it is becoming more difficult and expensive to farm it because the agricultural groundwork is not set up.

Increasingly threatened water supply. Lack of water may be an even more serious problem than limited or disappearing farmland. Fresh water for farming—already in short supply—is further reduced by silt deposits, pollution, and the sinking levels of underground water tables.

Energy. Petroleum is a major component in food production and distribution. Even though it seems to be in great abundance throughout the world, the price of petroleum remains at a much higher level than it was before the early 1970s. It is still high for developing countries that do not produce oil and must obtain it elsewhere for both industry and farming.

ecological having to do with the relationships between living things and their surroundings

Environmental corruption. Many of the world's poorest countries are under severe **ecological** stress, made worse by the growing misuse of agricultural resources, energy, and cattle feed. This reduces farm productivity in these regions, and the world is threatened with permanent environmental losses.

Technology shortage. Technological progress has contributed greatly to the well-being of all people and to the improvement of food systems. In spite of the great improvements in modern technology that have increased the world's production of food, the number of hungry people has not dropped. The misuse of some

▶ Children strip wood from a small tree on a plain devastated by years of drought (Mali, Africa).

expensive technologies (such as machines, irrigation, and chemicals) has worsened the nutritional situations of the poor in developing countries by taking away both their basic diet and their jobs.

Incomplete research. Farm research for developing countries has greatly boosted food production. But little research money is spent on tropical agriculture, food crops for local use (rather than for export), the study of how the different types of production affect nutrition, or the development of techniques to help small farmers and the rural poor produce more needed food.

Unpredictable weather. Since 85 percent of the world's farm products are grown on land that receives rain, the most crucial, long-term factor is the weather, which is neither predictable nor controllable.

It is possible to reduce some of these problems: (1) substitutes can be found for fossil fuels, such as oil and coal, and energy-saving techniques can conserve them; (2) trading and stockpiling more farm-grown goods can guard against the effects of bad weather conditions; and (3) improved farming practices can increase crop yields and conserve moisture and soil resources. But it is very difficult to deal with all of the problems at once.

The Demand Side

Although there are some difficult ethical issues on the supply side of this problem, those on the demand side may be even more difficult.

Population growth. The enormous increase in the number of the world's people puts great stress on the global food supply. As population grows, more food must be produced. History suggests that population growth will hold steady or increase in areas where there is much poverty and hunger. People in very poor countries

may have large numbers of children because they want to ensure that some children survive to work the farm, carry on the family and its culture, and help older family members. Population growth is not the cause of hunger. In fact, both population growth and hunger are symptoms of poverty and underdevelopment.

Rising wealth. Overuse by the rich causes great strain on resources. This overuse places far greater pressure on global resources than any use by the poor. Surveys show that the first increase in wages often goes toward trying to improve a family's diet. In most cases, this means shifting from eating foods such as potatoes and turnips to eating grains, and then from eating grains to eating poultry and meat. This dietary change places greater stress on the grain supply because between two and seven pounds of grain are required to produce a pound of chicken, pork, or beef.

Unjust social and political structures. Hunger is a symptom of the social "diseases" of underdevelopment and injustice. People are hungry because they are poor, and they become poor and stay poor because they don't have the power to choose otherwise. If people have the option, they generally will not choose leaders or policies that make or keep them poor.

Approaches to Ethical Analysis of Food Policy

As late as the 1970s, Anglo-American moral and political beliefs did not justify providing aid to undernourished people in foreign countries. Human-rights supporters were in favor of traditional Western civil and political freedoms but stood against **substantive rights** that required government involvement.

A new look at the issue of world hunger in the 1970s and 1980s resulted in a number of important changes in beliefs concerning both human rights and the interests of the needy. Philosophers began to turn their backs on traditional positions. They objected to government assistance for deprived individuals. By the mid-1990s, moral philosophers came up with three views concerning world hunger and international food policy: **subsistence** rights and duties, global justice and development, and humanitarian aid.

Subsistence rights and duties. Advocates for subsistence rights argue that some goods are necessary for the enjoyment of any other rights and so become basic rights themselves. Personal security would be one such right. In a similar way, the right to subsistence qualifies as a basic right because without the right to food, people will fail to enjoy any of their other rights.

Along with the right to subsistence come certain duties to which governments and citizens of all countries are morally obliged: (1) the duty not to deprive others of what is rightfully theirs; (2) the duty to prevent others from being deprived; and (3) the duty to help the hungry.

In a May 28, 1998, press conference, Dr. Jacques Diouf, director-general of the UN Food and Agriculture Organization (FAO), announced that global staple food production had shown a small increase for 1997–1998 but had in fact declined in eighty-three of the world's poorest countries. In many cases, especially in Africa, these countries cannot produce enough food to meet all their needs and lack the foreign exchange to pay for food on international markets.

substantive rights rights such as life, liberty, property, or reputation that make up part of the normal legal order of society

subsistence the minimum amount of food needed to stay alive

Global justice and development. The duty not to deprive is similar to the rule "do no harm." The duty to prevent others from being deprived includes developing cooperative organizations that will work to avert threats to food supplies. Such threats would include drought cycles, cutting back of irrigation techniques, and improper farming techniques. The duty to help the hungry includes providing emergency assistance to crisis victims.

The duty to prevent hunger has required reform of worldwide economic and political institutions. These institutions have had to rethink how the principles of justice apply to relationships between nations. In the past, international relations caused serious problems for those trying to work for justice between nations. The principle of "home rule" required governments to be completely responsible for their own people.

Theories based on the universal rights of individuals formed the basis for the principles of international justice. According to these theories, the increasing mutual economic dependence of nations sets up conditions for changing the ideas of home rule and **nonintervention** to promote justice between nations.

nonintervention the policy of not getting involved in the affairs of other countries

FAO workers build terraces in a South Korean village. This is a joint project of the United Nations and the government of South Korea to increase agriculture production through efficient use of fertilizers and by bringing wasteland into cultivation.

Some supporters of global justice in food policy argue that natural resources are distributed randomly throughout the world. These resources do not "belong" only to those nations in which they are found. Because these nations have done nothing morally deserving of these resources, they should not hold exclusive control of the resources or profit from them when people of other countries are facing hunger.

> ### FAO
> ### Food and Agriculture Organization of the United Nations
>
> The Food and Agriculture Organization was founded in October 1945 as one of the agencies under the newly created United Nations. Its mandate was to raise levels of nutrition and standards of living, to improve agricultural activity, and to better the overall conditions of rural populations.
>
> FAO is the largest autonomous agency within the United Nations with 174 member nations, a staff of 1,500 professionals, a budget of $650 million, and billion-dollar investments in agriculture and rural development projects.
>
> Since its creation, FAO has worked to alleviate poverty and hunger by promoting agricultural development, better nutrition, and the pursuit of food security, which is the access of all people at all time to the food they need.
>
> FAO is active in land and water development, plant and animal production, forestry, fisheries, economic and social policy, investment, nutrition, and food standards. It also plays a crucial role in dealing with food and agricultural emergencies.
>
> One of the agency's specific missions is to encourage sustainable agriculture and rural development, a long-term strategy for the conservation and management of natural resources. Such a strategy implies meeting the needs of both present and future generations through programs that do not damage the environment and that are technically appropriate, economically viable, and socially acceptable.

Those who disagree with this view believe that redistributing the wealth from nations of the North to nations of the South does not ensure a fair distribution of wealth within poor countries. It also **destabilizes** the necessary order in the international system.

destabilize to cause to be nonfunctioning

Humanitarian aid. Until the dawn of subsistence rights and global-justice theories, emergency food relief was given as humanitarian aid. This continues to be the case among voluntary agencies that provide emergency relief. Traditionally, **humanitarianism** has made two assumptions: (1) provision of emergency aid is an activity outside the normal boundaries of international politics, and (2) bringing relief to those suffering from a crisis, such as an earthquake, is a matter of doing more than is required under the law, based on motives of charity or compassion.

humanitarianism the promotion of human welfare and social reform

Difficulties in reaching victims of starvation in situations of civil conflict have raised two specific issues regarding humanitarian aid: (1) donors' right of access to populations affected by civil conflict, and (2) the duty of "humanitarian intervention" by the international community.

1. *Right of access.* In the 1990s, private agencies and government offices argued for greater freedom for humanitarian activities. An important step in this direction was a movement to grant relief agencies access to people caught in areas of civil conflict. While keeping a separation between humanitarian aid and international justice, such action appeared to insist on the rights of individuals over states, which gave human rights a new political prominence.

2. *Humanitarian intervention.* The idea of humanitarian intervention when there have been serious violations of civilian rights by military forces has been a part of just-war theory since the mid-1970s. The Gulf War (1991) and the crises in Somalia and Bosnia-Herzegovina (1992–1994) provided new meanings for the term. "Humanitarian intervention" has characterized missions of mercy by international organizations such as the Red Cross, and governments that are not part of the conflicts, when helping afflicted civilians in areas of civil unrest.

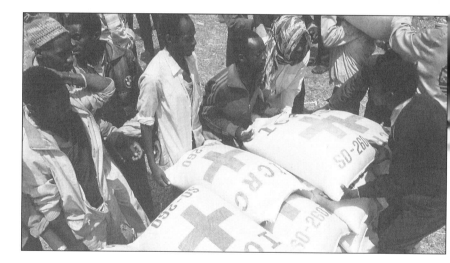

The Red Cross distributes food in Zaire (1994).

Ethical Questions

The universal existence of hunger and malnutrition reflects not only production shortfalls, lack of a decent food supply, poor technology, and climatic changes but also basic economic, social, and political structures. Programs that concentrate on food alone will not solve the problem of food security. Political and social barriers to improving food availability often are even harder to deal with than technical obstacles. They involve more uncertainty and are more resistant to change.

Over the long term, demand-side considerations may bring up more ethical questions than those on the supply side. In other words, supply-side questions are mostly technical and related to production, and therefore generally get economic answers. On the

At the World Food Summit of 1996, 186 countries committed themselves to reducing by half the number of undernourished people by 2015. According to 1998 United Nations figures, more than 800 million people in the world's 83 lowest-income nations are undernourished.

demand side, the problems (except for population growth) are more likely to be more political: Who gets what? Under what conditions? And who decides?

Should food aid be a part of foreign policy? Although foreign-aid programs have always had political as well as humanitarian motivations, the ethical answer to the question has to be no. The question really is whether food, which sustains life, should be treated like every other resource. Food aid from the United States has regularly been provided directly to governments through loans. But donated food has generally been given to the UN World Food Program or to organizations not connected with the government. The question of how much either of these methods leads to the recipient's dependency on aid remains unanswered.

▶ Lunch is served at a community center in Afghanistan. The project is sponsored by the World Food Program.

Should food relief today be limited because of more serious problems tomorrow? Some ethicists argue that food aid does more damage than good because it encourages the poorer nations to postpone improving their own food systems. It also allows the present generation to live and add more people to be fed in the future. The ethical results of this are clearly a problem.

Can some nations be classified as "nonessential"? Harsh as this question may seem, hard choices are made every day in many policy areas, including food policy—the Middle East versus Africa, Southeast Asia versus Central America, Eastern Europe versus developing countries. Some ethicists argue that guidelines for such decisions should consider human needs above all else. Racial, ethnic, cultural, and religious factors that might be considered in the decision should also be ethically evaluated.

Is there a duty to reduce consumption in wealthier societies? The widening gap between rich and poor nations (and between

 Researchers at an Indian university select high-yield corn seeds to plant 7 million acres of land. This large-scale effort is part of a project for which the Indian government received a $13 million grant from the World Bank.

rich and poor within nations) presents a strong ethical challenge. The buying and eating of grain-fed beef and pork, the wasting of food, and the overuse of energy can strongly affect both the availability and the price of food all over the world. Some would argue that simpler lifestyles in the rich nations have become a moral necessity in order to make equal distribution of the world's resources possible. Others respond that this lifestyle change would have little or no effect on food distribution and that no one has a right to what another has lawfully earned.

Conclusion: Beyond Food Policy

The global food system has not succeeded in getting everybody fed, and the market alone will not distribute food equally. In order to understand the basic food problem fully, it is necessary to look closely at investments, trade, finance, and aid. This requires examining a global system run mainly by its most powerful members—international corporations and financial institutions—in their own interests. Governments generally occupy no more than second place in the global market. Although these corporations and banks may eventually consider help to the poor to be in their interest, so far they have not.

Because the world's ongoing food crisis will remain a major problem, the ethical issues raised by it are likely to face both decision makers and ordinary people. But the food crisis, together with the environmental crisis, can neither be discussed intelligently nor resolved reasonably without attention to public-policy matters. These matters include corporate business practices, bank loans, population programs, social and economic development plans, foreign assistance, trade and investment, and social values and priorities. The position one takes toward the problem of feeding the hungry will be part of one's overall social ethical attitude. This position relates to a more general understanding of the rich's obligation to the poor, the government's role in providing assistance, and the role of economic factors in human development.

Related Literature

In 1906, Upton Sinclair published his muckraking novel *The Jungle*, which exposed the injustices and unsafe processes in the meatpacking industry. The industry had little concern for cleanliness. Laborers worked in cold and wet environments in the winter, and hot, fly-infested rooms in the summer. Immigrant workers especially suffered from the corruption in the industry and the sickening conditions of the workplace. Sinclair's powerful criticism led to many reforms.

POPULATION ETHICS AND POPULATION POLICIES

This entry consists of three articles explaining various aspects of this topic:

Elements of Population Ethics
Setting Population Limits
Strategies of Fertility Control

ELEMENTS OF POPULATION ETHICS Population studies deal with fertility, mortality, and migration. Fertility refers to births, mortality refers to deaths, and migration refers to the movement of people from one region to another.

Definition of Population Ethics

Population ethics has two main foundations: moral principles and facts. Moral principles come from religious traditions, philosophy, declarations of human rights, and other sources. Facts are learned through careful study of what is happening or has happened in a particular place or situation. Judgments about what is right in population policies require that moral principles—based on facts—be applied to cases.

The moral principles that guide discussions of population problems include preventing environmental pollution, keeping the number of people low enough so that there is enough food and space for everyone, and promoting economic growth.

The Population Problem

Newscasters, reporters, and politicians often talk about "the population problem." This phrase usually means that if there are too many people living on the planet, this will cause problems for individuals, couples, families, countries, or even the entire world. "The population problem" can also mean that a country or region has too *few* people for its own economic, social, or political good.

Every time people talk about a population problem, they are expressing concern about values such as preventing **famine**, having enough workers and jobs, and giving couples the choice of how many children to have. Whether the concern is that there are too many or too few people in a region, those who believe there is a problem always mention some moral, social, or political principle. One person who says there are too many people may mean that there is not enough food for them all. Someone else may mean that the large number of people is affecting the quality of the environment. Yet another person may believe that there is enough food, and the environment is all right, but that because things are more crowded and resources are limited, people have an average "quality of life" instead of a high one. One person may believe a region is crowded, while another person may not see it that way.

famine a widespread lack of food that causes many to starve

Projections of Population Increase

According to the Department of Economic and Social Affairs of the United Nations' Secretariat, the world population could grow in very different proportions, depending on how nations manage to control their fertility and birth rates:

- Assuming a medium-fertility scenario where fertility stabilizes at slightly above two children per woman, the world population will grow from–

  5.7 billion persons in 1995 to   9.4 billions in 2050

                                   10.4 billions in 2100

                                   10.5 billions by 2150

                                   11.0 billions around 2200

- With a lower fertility rate of one child per couple, the size of the world population would be 3.6 billion persons.

- A high fertility scenario, with one more child per couple above the replacement rate, would make for a population of 27.0 billion persons by 2150.

- If fertility rates were to stay constant at 1990-1995 levels for roughly the next 150 years, the world in 2150 would need to support 296 billion persons.

- The ultimate population size of nearly 11 billion persons, based on a medium fertility scenario, is 0.7 billion persons fewer than previously announced by the United Nations in 1992, mainly due to larger-than-expected declines in fertility in many countries.

Confusion about whether there is a population problem comes up when critics are vague about what they mean by "population problem." Are they measuring the problem using their own values about quality of life or what is right? What conditions do they think it will affect? Can they provide facts to back up their statements? Some writers simply take for granted that the world is too crowded and then go on to say what should be done about it. But if these writers do not give facts to back up their statements, others may say they are biased and may not pay them any attention.

A street scene in Bombay, India. Overpopulation creates problems of food production, housing, sanitation, health, education, and employment.

Mothers and children receive a bowl of soup as part of a food program supported by the World Health Organization and UNICEF (Hong Kong, 1960).

global warming an increase in the average surface temperature of a planet (Earth); also called the greenhouse effect

see also

Environment: Climatic Change

The Commission on Population Growth and the American Future was appointed by President Richard M. Nixon in the early 1970s to study the population problem in the United States. The commission concluded that gradual stabilization of the American population is a useful goal, with the birthrate roughly equal to the death rate.

per capita for each person

Approaches to the Population Problem

In his book *The Population Bomb* (1971), Paul Ehrlich wrote:

> The battle to feed all of humanity is over. In the 1970s and 1980s millions of people will starve to death. . . . Although many lives could be saved through dramatic programs to "stretch" the carrying capacity of the earth by increasing food production and providing for more equitable distribution of whatever food is available . . . these programs will only provide a stay of execution unless they are accompanied by determined and successful efforts at population control.

During the 1970s and 1980s high birthrates did not produce the levels of starvation that Ehrlich predicted. This was partly because of the Green Revolution, which was a series of technological and agricultural advances that led to much greater food production than in the 1960s. Nevertheless, in their 1990 book *The Population Explosion*, Paul and Anne Ehrlich continued to argue that the human species would face starvation and widespread disease unless societies immediately controlled their birthrates.

The Ehrlichs claimed that overpopulation caused starvation in Africa, homelessness and drug abuse in the United States, **global warming**, holes in the atmosphere's ozone layer, fires in tropical forests, garbage-strewn beaches, and drought-stricken farms.

The World Bank took a different approach to the population problem. Its *World Development Report 1984* agreed that the evidence on overpopulation is complex. But the report concluded that "population growth at the rapid rates common in most of the developing world slows development." What this means is that in developing countries, high fertility and fast population growth cause problems by creating other conditions—like lower-quality education—that block economic development. The World Bank believes that what governments do to change the rate of population growth will affect the world economy, and conditions in developing countries, well into the future.

In 1971, the National Academy of Sciences (NAS) claimed that rapid population growth causes serious harm to economic development in various ways. It holds down growth in **per capita** income, leads to unemployment, creates mass poverty, distorts international trade, and provokes conflicts between groups with different languages, ethnic backgrounds, religions, and political views. It slows the mental and physical growth of children, and has other negative effects.

Over the next fifteen years, the NAS shifted from a negative view of population growth to a more neutral one. In 1986, the National Research Council of the NAS issued a report that backed

away from earlier statements. According to the 1986 report, slower population growth may help developing countries, but it is difficult to know if this effect will be large or small. Also, the results of population growth do not depend only on the number of people in a region. The results can also depend on the effectiveness of government administration, social institutions, and the resources of the countries involved.

Population Growth by Continent

Population growth is far from homogenous, and projections only confirm a long-established pattern. Based on the United Nations' scenario of two children per family, population growth will continue in all major areas of the world except Europe.

	1995	2150
Africa	0.7 billions	2.8 billions
Asia	3.4	6.1
Latin America	0.4	0.9
North America	0.3	0.4
Oceania	0.02	0.05

Europe's population stood at 728 million persons in 1995, and it is projected to fall to 595 millions by 2150, a decline of 18 percent over 155 years.

In his book *Population Matters: People, Resources, Environment, and Immigration* (1990), Julian Simon provided a much more optimistic view of population growth than did the Ehrlichs, the World Bank, and the NAS. First, he questioned what he called myths about populations and resources. While others said that more and more people in developing countries did not have enough food, Simon believed that per capita food production had been increasing about 1 percent each year. Simon also did not believe that higher population growth means lower per capita economic growth. He also gave evidence to challenge the idea that the world is running out of natural resources and raw materials and that energy is becoming more scarce. In addition, Simon argued that having additional children improves productivity in the more developed countries and raises the standard of living in less developed countries.

These four approaches—that of the Ehrlichs, the World Bank, the NAS, and Simon—all had different notions of how population growth affects economies and societies. The Ehrlichs were continually dispirited about the effect of population growth on human societies. The World Bank was seriously concerned about its effects,

In Beijing, China, a caretaker pushes a wooden "stroller" with five toddlers. China's population is more than 1.2 billion.

and generally negative in its conclusions, but it was willing to consider different points of view as well as evidence that challenged its position. Like the World Bank, the NAS focused on population growth and economic development, but came to very different conclusions in its 1971 and 1986 reports. Simon played down the harms and said that population growth may actually have advantages and may lead to economic development and improved social welfare.

Is There a Population Problem?

If we focus on one country at a time, use information from the Population Policy Data Bank at the United Nations, and define "population problem" as the government's opinion on whether it has too few, too many, or the right number of people, it is possible to answer that question. But when we look at the world as a whole, and when we use different values, different theories about population, and different sources of evidence, then it becomes impossible to come up with one final answer.

If critics are going to talk about a population problem, they need to be sure they are all talking about the same thing. Many discussions of population problems are based on people's opinions about the state of the world, but are not based on facts. This must change in order for people to have more accurate, less biased, and more comparable discussions of population problems.

Thomas Malthus and his followers saw a growing population as a major threat, with starvation around the corner. Marxists believe that it is not population but the greed of the rich countries who do not want to share their wealth that is the real problem.

Malthus

Thomas Robert Malthus (1766–1834) was England's first authority in the new discipline of political economy. He spent most of his life collecting data on the relationship between population and its social, economic, and natural environments. He based his theory on these facts and adjusted it when others criticized it. His *Essay on the Principle of Population* (1798) went through seven editions.

According to the principle of population as explained in the *Essay*, population, if left uncontrolled, doubles once every generation. According to Malthus, this happens among nonhuman animals because they lack the capacity to think and to control their reproduction. Animal population drops when overcrowding leads to disease or starvation.

Because human beings are rational, they can see overpopulation's dangers and can control their natural drive to reproduce. There are two types of control of human population growth: "preventive checks," such as people waiting to have children until after they are married and able to take care of them; and "positive checks," such as people dying from sickness and hunger caused by overpopulation.

Malthus believed that the tension between the number of people and the available resources could have a beneficial effect. A

Thomas Robert Malthus (1766–1834).

man who cannot marry until he is able to support a family has an incentive to work hard, and his hard work benefits everyone in society. For this reason, Malthus opposed the use of contraceptives. He believed that if people could have sexual fulfillment without having to work to provide for the children they might produce, they would not work. This would be of no benefit to society.

In each edition of his *Essay*, Malthus increasingly stressed the negative connection between family size and social class. In his time, poorer people had larger families. In his view, this was the key to solving what in later times would be called the population problem. Malthus believed that the upper classes had more self-control and social responsibility than the lower ones had. He argued that the poor should be given more money and education, so they would have the same self-control. Malthus wrote that the main factors that led people to have fewer children were freedom, economic security, education, and a taste for life's comforts. Malthus's idea that increasing economic status leads to a decline in fertility is much less familiar than his ideas relating population growth to food, but it has turned out to be his most important contribution to population theory. Most of the populations that Malthus discussed tended to grow too rapidly in terms of the resources available, and he recommended official policies that would reduce their fertility.

> "Since it is the poor and disadvantaged who are most directly affected by the degradation of the natural environment, as resources become scarce and the quality of the environment declines still further, even more people are bound to become poor and disadvantaged. The best hope of limiting the increase in the number of such people would be if the world population could be stabilized."
>
> —Philip, Duke of Edinburgh (Prince Consort of Queen Elizabeth II of England), President, World Wide Fund for Nature, 1990.

Do families have a right to determine their size? The UN Declaration on Populations (1967) states: "The opportunity to decide the number and spacing of children is a basic human right, and family planning, by assuring greater opportunity to each person, frees man to attain his individual dignity and reach his full potential." Thus, the United Nations affirmed freedom of choice, but also strongly recommended that contraception be made available to families so that they can exert greater control over the numbers and spacing of pregnancies.

> "It is a country's sovereign right to decide its own population policy."
>
> —Li Zhaoxing, spokesman for the Chinese government, in response to U.S. criticism of alleged forced abortions as part of China's family planning program (1989).

SETTING POPULATION LIMITS Population policies raise profound ethical questions. Answers to such questions require us to clarify our ethical principles and to apply them to population problems across all countries and cultures. Our ideas about what is acceptable are shaped by the culture in which we live. Principles that make sense to Americans or Europeans may not make sense to people in other countries. An urban American may tell a rural Indonesian that it is not a good idea to have too many children. But the Indonesian may regard each child as a treasure, and say that she cannot work the family farm without her children.

Is China justified in using coercion to enforce its policy of one child per couple? (People who have more than one are fined, lose certain rights and privileges, and may suffer worse punishments.) Is it right for government officials and community members in Indonesia to pressure people to use birth control? Should U.S. judges be allowed to order surgically implanted birth-control devices for women they consider unfit to be mothers? Do the wealthy nations of the world have a moral duty to accept refugees from poor countries?

▶ Homeless Indians on a street in Calcutta. India's population is fast approaching the 1 billion mark.

Opinions on Population Control

Three schools of thought have guided debates on the ethics of population principles. The first argues that government programs of any kind must respect human rights as stated in the Universal Declaration of Human Rights adopted by the United Nations in 1948; the International Covenant on Economic, Social, and Cultural Rights (1976); the International Covenant on Civil and Political Rights (1976); and many related UN statements.

A second school of thought holds that the morality of population-control methods must be determined by the country that uses them. Each country understands its own problems and its own culture best, and can come up with the best way to control population. This school does not accept universal standards of human rights. It believes that when outsiders attempt to impose these standards on another culture, this interferes with each country's right to determine its own policies.

The third school recognizes some or all of the rights affirmed by the United Nations. But it claims that when overcrowding creates major economic or social problems for a country, its government has the right to limit individual reproductive freedom in order for conditions to be better for everyone.

In 1945, political scientist Frank Lorimer proposed three definitions of overpopulation. An area is overpopulated if:

1. a smaller number of people would lead to a high standard of living;

2. population increase is more rapid than possible production increase;

3. continuing growth trends jeopardize economic advances.

Five Key Principles

Making ethical evaluations of population policies requires five principles to guide decisions:

- life
- freedom
- welfare
- fairness
- truth-telling

It also requires that we have standards for determining when one principle is more important than another.

Life is at the top of the list, obviously, since without it people cannot benefit from any of the other four principles. Article 3 of the Universal Declaration of Human Rights states: "Everyone has the right to life, liberty and security of person."

Life means not only being alive but also enjoying good health and reasonable security against the actions of others that might cause death, illness, severe pain, or disability. Policies on fertility control, migration, and refugees threaten this principle when they take no action to assist people who are facing starvation or murder, or when they unwittingly create incentives for people to kill female babies. China's policy that each couple have only one child, combined with Chinese cultural beliefs that value sons more than daughters, leads some couples there to kill or neglect female babies, and to keep trying until they have a son. The one-child policy is not inherently inhumane, but it leads to inhumane conditions. Policies can also endanger health when they promote methods of fertility control that involve great risks to people's health. These methods include sterilization, oral contraceptives, intrauterine devices, or injections.

Freedom is the ability and opportunity to make thoughtful choices and to act on those choices. Freedom requires that people know about the choices that are available, such as options for fertility control or migration. People should also be able to make choices without strong pressure from others. They should consider the issues that are at stake and understand the possibility of taking action to carry out their choices. If any of these conditions are restricted, freedom is limited or eliminated. People who do not know what the options are, who are tortured, or who are not allowed to make choices, are not free.

UN statements are strongly in favor of freedom. According to the Universal Declaration, everyone has the right to freedom of thought, conscience, and religion; freedom of opinion and expression; freedom of peaceful assembly and association; freedom from slavery and servitude; and freedom from interference with privacy, family, and home.

This is what Articles 25 and 26 of the Universal Declaration state, as they relate to issues of population policies:

Article 25: "Everyone has the right to a standard of living adequate for the health and well-being of himself and of his family, including food, clothing, housing and medical care and necessary social services. . . . Motherhood and childhood are entitled to special care and assistance. All children, whether born in or out of wedlock, shall enjoy the same special protection."

Article 26: "Everyone has the right to education."

Welfare means a standard of living that includes adequate food, clothing, housing, health care, and education. The World Population Plan of Action of 1974 clearly connected population policies to human welfare. This document stated that "the principal aim of social, economic, and cultural development, of which population goals and policies are integral parts, is to improve levels of living and the quality of life of the people." Population programs should not aim only to raise or lower fertility, reduce mortality, or control migration. They should also promote human welfare.

Fairness refers to an equal and just distribution of the benefits and harms from population policies. This does not mean that everyone has the same level of benefits and harms. It does mean that one individual or group should not have much greater advantages or disadvantages from a particular policy. The Universal Declaration is strongly in favor of fairness. In Article 1, it states: "All human beings are born free and equal in dignity and rights."

In 1972, Ugandan President Idi Amin Dada wanted to reduce the population of his country. He ordered over forty thousand Asians living in Uganda to leave his country. His action is an extreme example of the unfairness that happens when a single group must pay the costs of a population policy. Unfair population policy implementation also occurred in India between 1975 and 1977, when beggars and other poor people were forcibly sterilized. Other examples include testing contraceptives only on low-income women, even though they are designed to be used by all women, and not telling uneducated people who are going to be sterilized what the operation will involve. Often, people are not told that the operation means they will never be able to have children again, and they are not told what the medical risks and side effects may be. In each of these cases, the political, economic, social, and medical harms of population interventions fall more heavily on one group than another.

Truth-telling requires giving accurate information about population policies and avoiding lies, misrepresentations, distortions, and evasions about what the policy is, how it will be carried out, and what the results will be. Truth-telling is not explicitly stated in UN declarations of human rights, but it is a basis of the other four principles. Lies about policies of fertility control, migration, and refugees can put people's lives at risk when they involve possibly fatal consequences, such as dying from infections or being shot in enemy territory. They limit freedom by not giving people the information, such as the side effects of sterilization, that they need to make informed choices. Lies harm welfare when they cause risks to people's income, education, or job prospects. They violate fairness when they are more likely to be told to one group, such as the poor or an ethnic minority.

coercion the process of bringing something about by force or threat

Using Coercion

Governments apply physical force when they order armed police or military officers to take citizens against their will to clinics that perform abortions or sterilizations, or when they threaten to torture couples who have more than two children. They use severe deprivation when they demand that poor citizens be sterilized before they can get a job or receive food that they and their families need. They also use severe deprivation when they warn that parents with more than a certain number of children will be put in prison or have their houses torn down, or when they use other threats that involve serious risks to life, health, and welfare.

The one-child policy. China has relied on coercion to carry out its one-child-per-couple policy and limit its citizens' fertility. The Chinese government claims that its policies are voluntary, but its pressure on population workers to meet their target, especially in cities, has led to coercive acts. In 1979, campaigns for "voluntary" sterilizations and abortions were widespread in China, and the fine line between persuasion and coercion was frequently crossed.

Cultural bias against girls. China's use of coercion to reduce fertility has led people to kill female babies, or give them away for adoption. In traditional China, men had the basic duty of continuing their fathers' descent line by having a son. The boy could carry on the family name, support his parents in their old age, and inherit their property. Failure to have a son showed ungratefulness to one's ancestors and made men look weak in their communities. This tradition continues. If a man's only child is a daughter, he and his neighbors may feel that he has not fulfilled one of his most basic duties in life. Yet a successful one-child policy would mean that many males could not have a son. Analysis of this problem strongly suggests that there is a clash between a couple's normal desire to keep and raise their daughter and the limit on having a son that is imposed by the country's policies on fertility control.

China's coercive policies show the severe tension between limiting population for the common good and life, freedom, and fairness. If Chinese couples are killing female babies in order to try again and have a son, the one-child rule conflicts with the infant girl's right to life.

Government officials may say that they never meant to encourage the killing of female babies, but this does not take away their responsibility for the deaths that take place. A complete ethical analysis of policies must look not only at official declarations and events but also at the actions to which they lead. If the policy of one child per couple has led to child-killing, then by UN standards of human rights this killing cannot be justified by the struggle to control China's population. In most social policies, life has such a high value that it cannot be given away even for the strongest claims that it is for the common good.

A family planning poster in Chengdu, China, promotes the one-child policy with a smiling baby and doves.

Responding to the United States' criticism of his country's "one couple, one child" coercive policy, China's Foreign Minister in 1989, Qian Qichen, stated:

"If the U.S. population were five times its current size, it would be fairly easy for members of Congress to agree with China's family planning policy."

Violation of universal rights. Coercive policies also restrict human freedom in ways that cannot be justified. Unlike life, freedom can be and often is restricted for the common good. Laws, tax rules, and many other policies tell people what they should and should not do. But forcing people to be sterilized or have unwanted abortions, which has happened in China, violates the principles of freedom and human dignity in all the UN declarations of human rights. The moral question is not whether people should be totally free to set their family size. There is no country or culture where people are able to do this. The moral question is whether some limits on reproductive choice are a violation of human rights. Using force to promote smaller families definitely violates these rights.

China's population interventions also raise the question of fairness. Policies that lead to the abortion of female fetuses, the killing of female infants, or female adoption affect only girls. Because of abortion and child-killing, girls have a lower chance than boys of being born or of surviving to adulthood. If a girl is adopted, she will survive, but she does not have the opportunity to be raised by her birth parents. All three of these outcomes violate fairness by giving more benefits to boys than to girls and more harms to girls than to boys.

Stepladder Ethics: A Contrast

Ethical principles based on internationally accepted standards of human rights differ greatly from the stepladder ethics proposed by Bernard Berelson and Jonathan Lieberson in 1979. Berelson was president of the Population Council, a well-known center of research, training, and advocacy on population policy. Lieberson was a philosopher who served as advisor to the Population Council and taught at Columbia University.

Berelson and Lieberson believed that where possible, less severe population-control measures should be used. They wrote that the amount of coercive policy used should be directly related to the seriousness of the population problem, and that stronger methods should be used only if the less coercive ones have failed. They believed that violent or potentially dangerous coercion should not be used unless all other population control methods have failed.

Berelson and Lieberson's "moral stepladder" begins with voluntary policies and, if they fail, moves up the scale of pressure on people to the level of coercion justified by the seriousness of the problem. They did not mention fertility-control measures that involve threats to human life, but governments with severe population problems would be allowed to use these methods.

Summary

To be relevant to the hundreds of countries and cultures across the world, population ethics must be based on values that are widely

▲ A billboard in Peru promotes the idea that family planning makes for a better life.

shared. Principles based on the ideals or assumptions of only one society will often be rejected by people from other backgrounds. Also, people must know what principles are more important than others. People who write population policies must be able to answer one of the most challenging questions in ethics: Is it morally acceptable to sacrifice one principle, such as life, for another, such as the common welfare?

This article proposes four principles based on international declarations of human rights: life, freedom, welfare, and fairness. It adds truth-telling as a fifth principle that is important in itself and as a base for the other four principles. Life is considered the most important principle.

In contrast to stepladder ethics, which does not allow for human rights, the ethical framework discussed here is against any method of population control that has serious risks of death or that uses torture, slavery, servitude, or other degrading punishments.

If this framework is used, it would have the same advantages and limitations as all universal codes of human rights. The main advantage is that it can be used to educate policymakers and public educators about what is and is not morally acceptable in population programs.

When a program violates its standards, UN organizations, such as the Commission on Human Rights, and private groups, such as Amnesty International, could make a record of any human-rights abuses and could demand more humane policies. Already, universal codes have led people in various geographic regions, such as Europe or Latin America, to look at human rights from other perspectives.

A framework based on international human-rights standards gives no simple answers to the complex ethical problems of population programs. It does provide a foundation for people with different politics, religions, ethics, and cultures to talk about what is right and wrong in population policy. Without this foundation, there will never be any serious analysis or lasting agreement about what should and should not be done in population-control programs.

STRATEGIES OF FERTILITY CONTROL A shift in Western fertility began in France in the late eighteenth century. This shift, involving a reduction in family size, spread to northwestern and central Europe and English-speaking colonies by the second half of the nineteenth century. Changing attitudes, culture, social structures, and living conditions had all contributed to a lower birthrate. One way to reduce population growth in developing countries might be to change people's attitudes in similar ways. Educational and motivational campaigns might influence people's attitudes about family size, affect the age when people get married, or increase the use of birth control.

see also

Fertility and Reproduction: Fertility Control

Are these campaigns fair or are they an attack on traditional cultural values? If so, is this attack right or wrong? Are these programs telling people the truth about the advantages they are supposed to get from having smaller families?

People who work for these campaigns usually say that the attack on traditional cultural values is justified because the old values are restrictive or oppressive, and people will benefit from new attitudes. Campaigners also say that the advantages of decreasing population growth outweigh any damage that may be caused by changing the society's values or reducing its fertility. The ethics involved need to be questioned because the encouragement for these campaigns usually begins in the West and through international organizations, not in the targeted countries. This means that Western values may be imposed on people in developing countries.

Family-Planning Programs in Practice

In developing countries, pressure on individual women, families, or communities usually comes from three sources: family-planning workers (usually women), who go from door to door talking to people; messages in the media; and political leaders who may encourage couples to use family planning. Very little research has been done about what these people say. Most family-planning messages promise that people who have smaller families will have better lives than those with large families. Some messages promise that women will have better health and reassure them that birth control is safe and free of serious side effects.

Family-planning workers. Health workers consider themselves successful if they get more clients and convince them to keep practicing family planning. It is not important for the workers to hear the client's opinion, assess whether the client will really have a better life, or to tell the client what the real risks or side effects of birth control might be.

In better family-planning programs, workers help clients get medical advice or different birth control. They often provide some forms of birth control directly. Sometimes their jobs and their income depend on how successful they are persuading clients to use birth control.

For example, in Matlab, Bangladesh, the family-planning program recruited only educated young women who practiced family planning and who were married into the most powerful families of the area. The program leaders wanted these women to provide leadership to other young women, and make them think family planning was a good idea.

Media campaigns. Research shows that in developing countries, radio has been the most effective of all media for changing family-planning practice. Radios can be heard by women even when they are busy working, and can be understood by people who

---

> **Approaches to limiting population growth:**
>
> - education (information and persuasion)
> - ready access to birth control
> - incentives (including financial rewards) for smaller families
> - coercion

▶ A family-planning sign in a sports stadium in Bangalore, India.

Beyween 1988 and 1992, the Information, Education, and Communication (IEC) Division of the Nigerian Family Health Services Project was given the mission of increasing the use of family planning. To that end, IEC launched far-reaching national family-planning media campaigns. The division developed and produced radio, television, film, and folk programs in the major languages spoken in the country—English, Igbo, Yoruba, Pidgin, and Hausa. Nigeria's most popular singers sang lyrics that encouraged adults and youths to be sexually responsible. Within weeks of their release, these songs had reached the top ten of musical hits.

cannot read. Family-planning workers use the radio to spread the news that women who use birth control will be less poor and have better health, and that birth control is safe and available.

Television campaigns have been less successful in family-planning campaigns in developing countries, although some experimental programs have increased levels of family planning. Newspaper articles have not had as much effect as workers had hoped, possibly because more men than women read newspapers in these countries. In some societies, women's magazines have been more successful.

Political leadership. Firm moral statements by national leaders have changed some people's attitudes about population control. One example is the repeated statement by former president Ibrahim Babangida of Nigeria, "Four is enough." Most women in the country know this slogan, which seems to have been successful at least partly because government has encouraged women to decide how many children they want without asking their husbands or their husbands' relatives.

Family-planning programs are similar to all modernizing and technical-aid programs. They have little interest in preserving native cultures. They pay attention to the wishes of national governments, which they see as upholding the national interest. But they do not pay attention to minorities, or even majorities who want to keep their old values and customs. It should be obvious that family-planning programs are different from other aid programs. They attempt to change sexuality, reproduction, and fertility, which all have to do with family and marriage—matters that lie at the heart of most cultures.

Since the family in Western cultures has changed and become smaller without outside encouragement or help, why should

A poster at a Singapore family-planning clinic.

currently developing countries be treated any differently? Most population-control workers would say that it is unethical to stand in the way of development and not to try to close the gap between the West and the rest of the world.

Incentives

An incentive is something—such as a reward—that makes people want to act in a certain way or perform certain tasks. Many countries use incentives to encourage people not to have big families. Others use incentives to encourage people to have more children. In a way, any government policy that affects population-related behavior could be labeled an "incentive." The incentive offered is often money. In countries where the government wants people to have more children, it offers family allowances. In the late twentieth century, Japan offered a "reward" of 5,000 yen (about $38 U.S.) per month for each preschool-age child and 10,000 yen for a third child.

India is an example of a country that offers incentives for people not to have children. A family welfare program that began in 1985 offered women cash to come to a clinic where they learned about and were provided with birth control.

The Bangladesh government pays individuals who agree to sterilization. The payment is equal to a week's earnings for an unskilled rural laborer. The individuals are also given clothing. In China, couples may sign a contract with the government agreeing to have only one child. They receive a reward of a yearly cash bonus that may be as much as one third of the average worker's monthly pay, larger food allowances, longer maternity leave, more farmland, better housing, and special priority for admission to school, medical services, employment, and housing.

Ethical Dimensions of Population Incentives

Experts often judge the value of government incentive programs by comparing the economic *value* of preventing a birth with the economic cost of preventing that birth. In other words, if the cost of preventing the birth is less than the cost of raising the child, then, according to this theory, the incentive justifies the cost.

Utilitarianism. The ethical theory that most clearly supports this concept is called utilitarianism, which proposes that whatever results in the greatest good for the greatest number is right, regardless of its consequences for the individual. Although the definition of "good" does not work in economic and social terms, people often apply it this way, especially when they are using computerized cost-benefit analyses. Numbers, like economic costs, are much easier to analyze than other benefits that may be harder to count. Many of the people who design government incentives to change people's population-related behavior are trained in

economics. They may also look at the idea of social good purely in terms of money because of their training.

Individual rights. Western ethical theories have never completely accepted the idea that a policy is right if it produces the greatest good for the greatest number. The philosophical tradition of John Locke, Thomas Hobbes, and Jean-Jacques Rousseau recognizes that individual rights may conflict with something that is good for society as a whole. Immanuel Kant's tradition of philosophy, a major root of Western ethics, holds that actions or policies are right or wrong because of their built-in moral characteristics, not because of their consequences. When people decide whether government population incentives are right or wrong, they need to look at individual rights and the effect these policies have on these rights.

Evaluating incentive programs. Incentives may be a matter of free choice or people may be forced to accept them. This is complicated because any choice that people make about reproduction is affected by the society they live in, income levels, and opportunities for child care and education. Some incentives may not seem like coercion, but they are. In a country where people are starving, offering an amount of money that is many times more than a couple's annual income—or preventing them from getting food if they do not accept population restrictions—is coercion.

Some incentives that provide food, shelter, or clothing or that go against the usual pressures in the society may actually offer more choices. For example, if a woman lives in a society where there is great pressure to have a large family, but she does not want to have a big family, going against her family's and society's wishes may be difficult. If she is rewarded with money, housing, or clothing for having a smaller family, this may help her to live with her choice.

In a perfect world, no one would think government incentive programs are a good idea. Incentive programs are proposed only by those who think that otherwise people will continue to have large families. Incentives are also proposed by those who want some kind of government intervention in population but who do not like obvious coercion.

There are a limited number of ways to change populations without using incentives. Educational campaigns and basic social changes are the main alternatives to government incentive programs.

Summary. When evaluating whether incentive programs are right or wrong, we must take into account the particular country and culture that offers the incentive. We must also look at other possible methods that each country or culture can use to change its population. If other government policies will not work or will hurt people, incentives may be the only choice left to change population. But usefulness and efficiency are not the only principles to apply when deciding whether incentives are just. We must consider individual rights, freedom, and justice when evaluating incentive programs.

On August 1, 1972, new laws on birth control went into effect in Singapore. Maternity costs in public hospitals, already scaled to favor small families, were scheduled to vary even more pointedly. The cost of delivering a third child was pegged at $2\frac{1}{2}$ times that of a second. In its announcement, the government declared, "For the Singapore of the 1970s the third child is a luxury, the fourth and fifth are antisocial acts."

PUBLIC HEALTH

This entry consists of three articles explaining various aspects of this topic:

Introduction
Philosophy of Public Health
Public-Health Methods: Epidemiology and
    Biostatistics

INTRODUCTION Faith in medical science's ability to affect health is based mostly on the enormously successful experience with microbiology to fight disease. Science has identified specific microorganisms that are directly responsible for epidemics of certain **contagious** diseases as well as ways to control those microorganisms. The result is that many people expect science to find magic bullets for most of humanity's diseases. Other discoveries, such as insulin for diabetes and chemicals that fight certain forms of cancer, have strengthened the belief.

contagious easily spread by contact

The reliance on medicine tends to overlook far more basic influences on health. For thousands of years, it has been clear that living conditions and the response to them largely determine peoples' health. Therefore, people have tried to extend life and improve health not only as individuals but also through public efforts in their societies. These social efforts to improve the health of whole populations are what we call public health. Government plays the leading role in this mission, with additional support from other organizations dedicated to improving the health of the public. Making medical services available to people is only one way in which industrialized societies address health problems. Other steps include keeping the environment healthy and supporting healthful behavior.

Public health measures its progress by the health status of the population it serves. Thus, knowing what determines the public's health is an important part of the field. The most amazing health improvement in human history has occurred since 1800. During this time, **life expectancy** has grown much longer. Estimates between the time of the hunter-gatherers thousands of years ago and the Industrial Revolution around 1800 put life expectancy at birth as ranging between twenty and fifty years. Until 1800, most people lived for about twenty-five to thirty years. At the beginning of the twenty-first century, life expectancy is more than sixty-five years in most parts of the world and seventy-five years in western Europe, North America, and Japan.

life expectancy the average age to which most people will live

PHILOSOPHY OF PUBLIC HEALTH Public health is the prevention of disease and early death through organized community effort. While the government often leads this effort, many nongovernment organizations play major parts in improving the

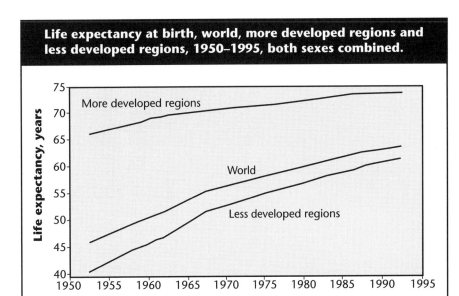

Life expectancy at birth, world, more developed regions and less developed regions, 1950–1995, both sexes combined.

publics health. Public health as an idea is one of the most important of our time. Since the middle of the nineteenth century, it has been a major force in changing the shape of the modern world and enlarging government's reach. The general idea that government and communities can choose to discover and relieve disease and social problems is relatively new in human history. It involves complex and related developments in the area of statistics and the study of different disease patterns in human societies (usually called epidemiology). It further requires a large, capable government to make use of these findings.

Public health's focus on populations and communities is the central source of its philosophical interest. The community outlook leads to a way of thinking about disease, early death, and their prevention that differs from many assumptions of modern bioethics and other areas of study. Public health as an organized practice views disease and early death from the point of view of the community and its ability to examine itself, reorganize, and change. The community outlook strengthens society's ability to discover the causes of disease in individuals and to come up with flexible and quick ways to control disease and preventable death.

Health by Design: The Idea of Prevention

The major focus of public health is prevention, which involves the creative planning of social environments and communities to better promote health and safety and replace older models of the problems that need to be solved. A major part of the battle in public health is redefining these problems in terms of the community's outlook rather than the individual's, so widely accepted in much of philosophy and social science, that it is a powerful obstacle to effective prevention. This is especially true in applying public-health

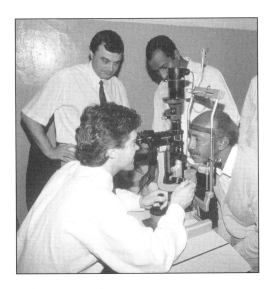

An Indian patient is having his eyes checked.

In 1907, Mary Mallon of New York City was discovered to be a typhoid carrier. She did not become ill herself but she infected people around her. Becoming known as Typhoid Mary, she refused all treatment and continued to work as a cook. Finally, in 1918, the public-health agency in New York put her in a hospital, where she was held against her will until she died twenty years later.

General Topics: Substance Abuse

methods to chronic disease, injury, alcoholism, and drug addiction.

In the case of alcohol, since the 1970s, there has been a shift away from purely individual explanations for alcohol problems. Individual explanations hinge on such factors as the abilities, nature, and motivations of people who drink, and who then have problems, including loss of control over drinking. The public-health focus is on the exposure of whole societies to alcohol, on the different levels of total drinking among groups, and on such factors as price, hours of sale, and age limits. This approach does not try so much to explain why some people become addicted to drinking, but why rates of alcoholism rise or fall among communities or over time.

In a similar way, highway safety since the early 1960s has witnessed a shift away from individual factors such as driver error, driver negligence, failure to yield the right-of-way, and factors beyond a driver's control. The emphasis now is on such factors as a driver's exposure to highway dangers, miles driven per year, types of roads driven on, and the safe or unsafe nature of the car. Exposure is an important factor in this redescription and often results in unexpected insights. For example, researchers have noted that driver-education programs in the United States probably raise the level of death and injury because they expose more young people to the hazards of driving at an early age.

In one way or another, public health concerns the choices a community makes. Public health is about how much alcohol to permit per person in society, about the frequency of highway crashes, and about the number of drownings in a state or nation. It is also about changes in environment, legislation, and public attitudes that will directly affect those statistics.

Fasten Your Seat Belt!

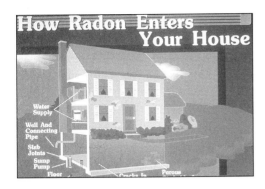

A chart created by the Environmental Protection Agency (EPA) showing how radon enters the house.

A woman smoking a cigarette pauses to read an anti-smoking poster.

General Topics: Substance Abuse

Public Health and the Common Good

The common good in public health means the good of individuals taken together as a group, as communities, or in terms of overall health and safety. This overall health, indicated by so many thousands of lives saved, is the goal of organized government or community effort. The common good does not mean that each individual has the same or identical good in health and safety, or even the same interests. A person with a genetic likelihood to get colon cancer does not have the same interest in health and safety as another who does not have such a genetic trait. Yet both can be said to have a common interest in promoting health and safety and reducing general risks to health and safety that all face, including cancer risks.

For most public-health problems, the overall savings in lives is far smaller than the number of people at risk and whose freedom is limited. In other words, the group that benefits from protections is just a small part of the larger group that is at risk. Thus, not all who are at risk and whose freedom is limited by public-health legislation will benefit. The benefit goes to an unknown minority of the larger at-risk group.

Democracy, Public Discussion, and Public Health

Much of public health is about providing information and education. These activities usually have far fewer ethical conflicts than do laws that limit individual freedom in order to improve health and safety. Yet, one can see the unique mark of public health as a social practice. Progress against cigarette smoking in the United States has been made not so much by banning smoking as by communicating public-health researchers' discovery of the links between smoking and disease. U.S. surgeons general prepared a series of reports that linked smoking to lung cancer and heart disease. These reports led to public discussion, making the link widely known. Discussion about the role of tobacco in public policy also created a growing awareness of smoking as a social problem. This awareness, together with the ban on television advertising of cigarettes, led to a sharp decrease in smoking rates even before later moves to ban smoking in public areas.

PUBLIC-HEALTH METHODS: EPIDEMIOLOGY AND BIOSTATISTICS Epidemiology is the study of the health of a population. It is basic to modern public health. It provides information that helps make decisions affecting health planning and how to spend money. It is the basis for deciding whether or not to introduce or change health policies aimed at preventing disease. Finally, it plays an important part in making decisions concerning the best way to treat diseases through testing the results of actual treatments.

Epidemiology is different from medical science because its focus is on the health of the population rather than that of the individual patient. Medicine seeks to heal the individual who, because he or she is vulnerable, becomes ill. Epidemiology tries to find the basic cause that results in illness among those who are vulnerable.

Epidemiology looks at large numbers of people that make up a population. It is a science whose methods rely heavily on using ideas about biological statistics and on advances in the methods of getting biological statistics. As with other sciences that depend on numbers and statistics, epidemiology requires the counting, sorting,and studying of large amounts of information. In order to get meaning from massive amounts of data, statistical methods are used to get different kinds of summaries. These methods are known as biostatistics.

Through the early 1940s, before the use of antibiotics at the time of World War II, epidemiologists tried almost exclusively to control **infectious** diseases. Success resulted in better control over those diseases, better living standards, and longer life expectancy of the population. Because of that success, epidemiology expanded from its concern with infectious diseases to include diseases that are **noninfectious.**

The idea of preventing noninfectious and chronic diseases by removing their causes is relatively new. Hence, the modern role of epidemiology is to find proper ways of prevention for policymakers to think about, in order to control disease where it starts and thereby improving the health of the community.

infectious spread by a virus or other microorganism

noninfectious pertaining to illnesses that cannot be spread from one person to the next

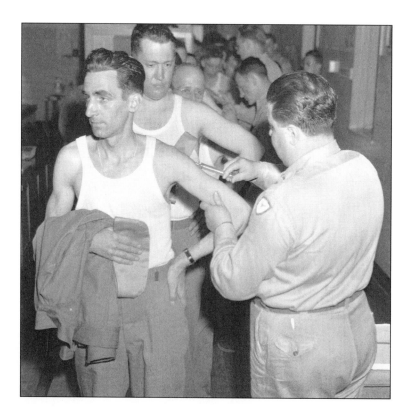

▶ A medic vaccinates a line of conscripts at a U.S. Army induction center during World War II.

The linking of epidemiology and biostatistics has become a symbol of modern epidemiology in both its research and its practice. Research in epidemiology tends toward experimental activities. The practice of epidemiology tends to focus on monitoring diseases. For both research and practice, biostatistics provide the tools used in epidemiology.

Ethics Guidelines

The first published, stated need for guidelines on the ethical conduct of epidemiologists appeared in 1985. Even though there was a lot of discussion and argument within the profession in North America, through 1987, there was little progress. Epidemiologists adopted the proposal to develop guidelines at the International Epidemiological Association's (IEA) 1987 Scientific Meeting in Helsinki, Finland. By 1990, further discussion had advanced the thinking on this subject and a first draft of IEA guidelines was published.

A landmark meeting on the subject of ethics in epidemiology caused discussion in 1989. The United States Industrial Epidemiology Forum organized the meeting. The Forum's Guidelines are organized as follows:

I. *Obligations to the subjects of research*

- to protect their welfare, ensuring no physical or mental harm through their participation
- to obtain their informed consent, ensuring the fullest possible understanding of any risks and benefits that may result from their participation
- to protect their privacy, ensuring no dishonor resulting from information provided through their participation
- to make sure that the participants and information about them stay private

II. *Obligations to society*

- to avoid conflicting interests, knowing that certain interests could influence research in ways that fail to serve the goal of finding the truth
- to avoid taking one side by openly stating which way one is leaning
- to widen the scope of epidemiology by teaching its methods to interested people
- to be as responsible as possible
- to make sure that the public trusts the profession by making known both the strengths as well as the limitations of the profession

Persons who are HIV-infected can spread the AIDS virus to others through sexual contact or blood products. Contrary to all other sexually transmitted diseases, HIV-AIDS does not have to be reported to the public-health authorities. This exception to the rule is highly controversial.

While guidelines, commentaries, and case studies are essential to ethical conduct, they are not enough. They must be taught, learned, discussed, challenged, and revised in light of case studies, if they are to affect conduct. Finally, methods for dealing with charges of unethical conduct should exist with remedies that serve to right any wrongs.

Related Literature

Henrik Ibsen, in his play *An Enemy of the People* (1882), portrays a public-spirited Dr. Stockman who discovers that the public baths for which the town is famous are contaminated, probably sickening the tourists who use them. As he tries to clean up the water supply, he runs into conflicts with political enemies who do not want the expense of rebuilding the town's water supply and who are not convinced that the doctor knows what he is talking about. The doctor's own uncompromising idealism and insistence on principles only make the conflict worse. The political powers call him an enemy of the people and essentially force him to give up.

Sinclair Lewis, in his satirical novel *Arrowsmith* (1925), portrays a doctor who goes through several changes in his perspectives and values. During his time as Director of Public Health, Arrowsmith alienates many by bluntly telling them that they endanger the community's health by keeping diseased dairy cattle. When he closes the infected dairies, he is forced out of town by political enemies. Arrowsmith later discovers a cure for bubonic plague. The institute where he works sends him to a tropical island to test his serum. Wanting a controlled study, Arrowsmith keeps the serum away from several people, but when his wife and good friend both die of the plague, he injects everyone and saves the population.

PUBLIC POLICY AND BIOETHICS

As issues of right and wrong in health care and medicine in the United States during the 1970s and 1980s became an area of great concern, many bioethics questions gained attention. During these two decades, discussion of bioethics issues occurred not only in classrooms or between doctors and patients but also in public debates and policy decisions. One of the best examples of this development of public policy on important bioethics issues is cardiopulmonary resuscitation (CPR).

CPR is a life-sustaining treatment used on patients who have suffered a heart attack or have stopped breathing. Originally, this technology was developed for and applied in a narrow range of

emergency medical cases, such as a drowning. It saved otherwise healthy people who had suffered a sudden heart attack or had stopped breathing. Later, CPR came to be used in a wider range of cases, including those in which a benefit to the patient was not clear. Eventually, CPR was used even more widely because the emergency conditions to which it responded—cardiac and respiratory arrest—made it impossible for medical personnel to take the time to think about whether or not to use it.

Reports of unethical and bad medical practices led many hospitals to develop formal policies to limit CPR. This interest in Do Not Resuscitate (DNR) orders led to scholarly studies about the use of CPR and DNR orders. The Joint Commission for the Accreditation of Hospitals required hospitals to have a policy regarding DNR. In this case, a public-policy response to an important ethical problem in the practice of medicine led to both a public and a professional response. A broader debate in bioethics has also focused on DNR policies. This heated debate centers on whether resuscitation should be an option to patients or their surrogates (those people speaking for patients) when its use would be ineffective or burdensome.

In other cases, public-policy initiatives have tried to increase the use of a desirable practice. A case in point is the Patient Self-Determination Act of 1991. This act requires federally funded hospitals to tell patients, upon admission, of their rights under state law to use **advance directives**. With these directives, patients can indicate whether or not to have life-support treatment if this should become necessary during their hospital stay.

The Role of Public-Policymaking Agencies in Bioethics

The first U.S. commission. In 1974, the U.S. Congress established the National Commission for the Protection of Biomedical and Behavioral Research. Two important factors led to the creation of this first public, national agency to shape bioethics thinking and practice in the United States. First, the character of biomedical research had changed significantly during the 1940s, 1950s, and 1960s. Before World War II, such research was usually carried out in small-scale settings in which medical researchers were trusted by their patients or subjects and by the community. But during and following World War II, the scale of this research expanded greatly as the public excitement about the potential benefits from medical research grew. As a result, researchers were no longer doctors caring for patients, and the well-known and trusted health practitioner was replaced by the cold and unknown investigator. Second, public concern over research abuses grew. The shocking abuses of human subjects by Nazi doctors during World War II had drawn public attention to these issues. In 1966, a member of the faculty of Harvard

Death and Dying: Euthanasia and Sustaining Life

advance directives "living wills" and durable powers of attorney for health care

The Tuskegee Study of Untreated Syphilis in the Negro Male, which lasted from 1932 to 1972, involved 600 black men, of whom 399 had syphilis and 201 did not. The participants signed up with the U.S. Public Health Service and were compensated with free meals, free medical care, and free burial expenses. The men suffering from the disease were told that they had "bad blood" and that they would receive treatment. In fact, they were given placebos and were denied treatment even after penicillin was released in 1947. By the time the study was uncovered (1972), 28 men had died of the disease, 100 were dead of related complications, 40 wives had been infected, and 19 children had contracted the disease at birth.

Ethics and Law: Information Disclosure, Truth-Telling, and Informed Consent

In 1996, President Bill Clinton created the National Bioethics Advisory Commission. As a first task, the commission was asked to produce a report on the ethics of human cloning, prompted by the cloning in Scotland of a sheep named Dolly.

Medical School, Henry Beecher, published an important article in the *New England Journal of Medicine*. In his article, Beecher described twenty-three instances of published medical research in which the treatment of human subjects was morally wrong and unethical. Around the same time, some glaring cases of research abuse received wide public attention. One of the worst cases was the Tuskeegee syphilis study. In this study, African-American men who were suffering from this terrible disease were not given treatment, so that doctors could observe the natural course of the illness.

The 1978 Belmont Report, prepared by the National Commission, had a tremendous impact on bioethics. This report addressed the moral principles underlying various research aspects. The National Commission stressed the moral principle of respect for individuals. This meant that people should be subjects in research only with their free and informed consent, and with confidentiality. The work of the National Commission continues to form the ethical basis for the federal government's supervision of research using human subjects.

The President's Commission. When the National Commission finished its work in 1979, Congress established the President's Commission. Because of the broader nature of the topics addressed by the President's Commission, its reports had many effects on bioethics. For example, its report *Defining Death* (1981) contributed to the great majority of states' adopting the same brain-death standard. Its report on informed consent, *Making Health Care Decisions* (1982), had a widespread impact on advancing the idea that doctors and patients share decisions about treatment. *Securing Access to Health Care* (1983) focused on the ethical problems surrounding the fact that more than 20 million Americans did not have health insurance. This report had less immediate impact than many others because solving the problem required an enormous amount of the government's money. This was a time when the emphasis of the Reagan administration was on cutting, not expanding, government social programs.

Other public-policy agencies. The National Commission and the President's Commission have been the two most important public-policy agencies in the United States that have focused on bioethics issues. A number of other public agencies, however, have also focused on these issues. Several states, including New Jersey and New York, have set up their own bioethics commissions.

A striking example of how bioethics in the United States has become an accepted part of the public domain and of government-sponsored research is the Human Genome Project. This $15 billion, fifteen-year project was established in the mid-1980s to map and sequence the genetic code in humans (the genome). At the time Congress debated the project, there was concern about its ethical, social, and legal effects. James Watson, the first director of the Center for Human Genome Research at the National Institutes of Health,

Genetics: Eugenics, Genetic Mapping and
Sequencing

committed the center to spending at least 3 percent of its total budget on research and education concerning these legal and bioethical issues.

Limits to public-agencies' recommendation. There is a difficulty in using government agencies to address divisive ethical and political issues. Consider the task force set up to study the use of tissue from fetuses in research. The task force's recommendations to permit limited use of fetal tissue were ignored by the Bush administration in the late 1980s because the use of tissue from an unborn child was so closely related to the controversial issue of abortion. "Right to Life" groups feared that any use of fetal tissue could increase, or appear to express approval for, abortions. The attempt by the U.S. Congress in the late 1980s to establish a bioethics committee to advise its Biomedical Ethics Board also failed in large part because of political struggles over abortion.

Conclusion

Bioethics issues have received a great deal of attention in public policy, and bioethics scholarship has strongly influenced public policy in health care. At the same time, public policy in the form of legal decisions and government public-policy agencies have deeply influenced the development of bioethics. As bioethics focuses more on broader issues of health policy in the future, this interaction and influence between public policy and bioethics is expected to increase.

Professional–Patient Issues

COMPETENCE

If a doctor is to treat a patient, the patient must consent to the treatment, and the patient must be competent to do so. Competent individuals may, in most situations, consent to or refuse health care. In contrast, a doctor does not need to obtain consent to treatment from a legally incompetent person, although a family member of the patient must usually grant consent. A mentally ill person who refuses to have a pacemaker inserted in her heart because she believes that Martians could then control her activities would not be considered legally competent to refuse medical treatment. Competence is usually not an issue in emergency situations because obtaining consent or refusal from the patient would delay the emergency treatment, which could be fatal to the patient.

Definitions

Generally, "competence" means the ability to perform a particular task. A person may be competent to perform one task but not another. A person may be competent at one time, but not at another. A person's competence to perform a certain task therefore depends on the situation, time, or place.

Competence relates to all areas of a person's functioning. Questions about competence are often raised about a person's ability to work, manage money, write a will, live alone, drive a car, marry and divorce, have a child, or testify in court.

In health care, "competence" means the capacity to make one's own health-care decisions. "Competence" is sometimes used in a narrow legal sense to mean that a court has ruled on a person's capacity. In this legal sense, people are considered **incapacitated** when they cannot perform a particular function, but considered incompetent only when a judge rules that they cannot perform that function. A person is considered legally competent until proved otherwise to a judge.

A person's capacity to perform a function or make a decision can change over time or depend on other factors. Also, a person's capacities cover a wide range. So, in most cases, a doctor does not have to accept that a patient is permanently incapacitated. Hearing aids, glasses, medication, counseling, and specific training in the area of the incapacity are examples of steps a person can take to improve capacity. When these efforts fail and it is determined that someone is completely incapacitated, the doctor must decide what to do, depending on the situation. If a patient is incapacitated and refuses lifesaving treatment, a doctor may have to get a judge to declare the patient legally incompetent so that the doctor can treat him.

incapacitated deprived of the power to function normally; legally not capable

The designation of a patient as incompetent or incapacitated should never be made lightly, since the freedom to direct one's own future is at stake. Patients have been unfairly declared incompetent. Some doctors even assume that the refusal of lifesaving treatment is always an indication of incompetence.

Competence Standards

There are no global standards for determining whether a person is morally or legally competent. In a given case, doctors, lawyers, and ethicists may agree that a particular person is incompetent in some respect. They may also disagree in many cases. Disagreement among professionals occurs because determinations are often not based on objective facts. Rather, they reflect the values of the doctor or other professional making the competence determination. These determinations may depend on how closely the patient's values conform to the doctor's. "Competence," then, is not a fixed characteristic but instead depends on the context and upon the professional making the determination.

Making choices. Most standards for competence determine whether the person being assessed can make a choice, can communicate that choice, can understand relevant information about the choice and its alternatives, and then can apply that information.

The sliding scale. In health care, the selection of competence standards depends on the decision at hand. In this "sliding-scale" approach, the competence standards vary with the particular decision and its risks and benefits. As the risks of the proposed health-care treatment increase (or as the benefits decrease), more capacity is required for the patient to be considered competent to consent to the treatment. It is less difficult to decide whether to take antibiotics to treat a case of the flu than it is to decide whether to undergo experimental treatment for cancer, so less capacity should be required to do so.

Although the sliding-scale approach is commonly used in health-care decision making, some problems appear with its use. Since most health professionals—and society—favor treating patients, one concern is that professionals will manipulate the competence standards to label someone who consents to treatment as competent and someone who refuses treatment as incompetent.

Assessment of competence. Doctors make informal judgments about a patient's competence daily. Some cases, however, such as treatment refusals by patients whose competence is unclear, require formal, detailed assessments. Competence assessments should focus on the particular function in question. These assessments sometimes involve observing a patient performing a function (like shopping or cooking a meal), psychological testing, or psychiatric interviews. In addition, the patient's family and friends, as well as other health-care personnel, may provide histories or reports about the patient. The doctor making the assessment tries to obtain information about the patient's decision-making history and the values he or she has placed on personal independence, health care, disability, and death. Typically, competence is not challenged, investigated, or formally assessed in clinical settings until a patient refuses or does not comply with treatment.

Members of the Jehovah's Witnesses oppose blood transfusions on religious grounds. Their right to refuse treatment has been upheld by the U.S. Supreme Court as protected by the First Amendment to the Constitution, which guarantees freedom of religion.

Health Care Proxy

(1) I, _____

hereby appoint _____
(name, home address and telephone number)
_____

as my health care agent to make any and all health care decisions for me, except to the extent that I state otherwise. This proxy shall take effect when and if I become unable to make my own health care decisions.

(2) Optional instructions: I, direct my agent to make health care decisions in accord with my wishes and limitations as stated below, or as he or she othwise knows. (Attach additional pages if necessary.)

_____
_____
_____

(Unless your agent knows your wishes about artificial nutrition and hydration (feeding tubes), your agent will not be allowed to make decisions about artificial nutrition and hydration. See instructions on reverse for samples of language you could use.)

(3) Name of substitute or fill-in agent if the person I appoint above is unable, unwilling or unavailable to act as my health care agent.

_____
(name, home address and telephone number)

(4) Unles I revoke it, this proxy shall remain in effect indefinitely, or until the date or conditions stated below. This proxy shall expire (specific date or conditions, if desired):

_____

(5) Signature _____

Address _____

Date _____

Statement by Witnesses (must be 18 or older)

I declare that the person who signed this document is personally known to me and appears to be of sound mind and acting of his or her own free will. He or she signed (or asked another to sign for him or her) this document in my presence.

Witness 1 _____

Address _____

Witness 2 _____

Address _____

Conclusion

Whether in health-care, financial, or legal decision making, the stakes for both individuals and professionals in defining competence are considerable. Identifying and labeling someone as incompetent can be humiliating and deprives the person of self-determination. On the one hand, legal and health-care systems face the need for **surrogate** decision making for the incapacitated or incompetent person. On the other hand, failure to protect the incapacitated person from making faulty or harmful decisions (such as refusing necessary medical care) may not honor the person's best interests. The question then is when and how to respect a person's decision.

In most health-care cases, doctors agree that the person should—or should not—be considered competent even if there is

surrogate a person who acts or makes decisions for another person, often to do with legal matters

no overall agreement on how much rationality and understanding are adequate for the person to be considered *legally* competent. Still, there are other cases in which judgments about the person's decision-making capacity are more difficult. In those instances, doctors, patients, families, and the courts become involved in emotional disputes over handling the person's medical care.

HOSPITAL

This entry consists of two articles explaining various aspects of this topic:

Medieval and Renaissance History
Modern History

▲ An Italian marble representing a woman providing treatment to a patient.

MEDIEVAL AND RENAISSANCE HISTORY The early history of hospitals dates from about 400 to 1600, and includes these developments:

1. the origins of hospitals

2. their development in the Byzantine and Islamic worlds

3. their history in medieval western Europe

4. their flourishing in Renaissance Italy

The term "hospital" refers to an institution that focuses on caring for patients and, if possible, curing them.

Modern hospitals trace their origins, and even their name, to the hospices and hospitals established by the Christian church during the late Roman Empire.

Early Hospitals

From its earliest days, Christianity demanded that its followers help sick and needy people. Christians believed that on the Last Day, God would judge each person according to the love that person had shown to others in need. Had one fed the hungry, sheltered the homeless, visited the sick? By the early second century, bishops expected their clergy to take care of the sick, orphans, and widows.

Hospices. Local Christian clergy helped the needy without any formal charitable institutions until the fourth century. After that time, in the eastern, Greek-speaking provinces of the Roman Empire, the demand for charity became so great, especially in the larger cities, that specialized institutions called hospices appeared. By the year 320 the church in Antioch operated a hospice to feed and shelter the poor of Syria. By the mid-fourth century, the pagan emperor Julian described hospices as common Christian institutions.

The first hospitals. Before 360, Christian hospices did not focus attention on the sick. But during the 370s Basil, bishop of Caesarea in Asia Minor, opened an institution where physicians and nurses treated patients. Twenty years later, Bishop John Chrysostom supervised hospitals in Constantinople where doctors took care of the sick. These early hospitals evolved from simpler hospices by expanding their services to include free medical care for needy guests.

Christian bishops built hospices and specialized hospitals for the sick during the fourth century. They built hospitals because they wished to follow Christ's command to practice charity and because they wanted to attract new Christians among the lower classes who lived in cities. Early hospitals received money from the income of farmland and other property that local bishops had given to support them.

Hospitals Begin to Treat Diseases

By the late sixth century, Christian hospitals such as the Sampson Hospital of Constantinople (modern-day Istanbul) maintained specialized wards for surgery patients and those with eye diseases. The best physicians of this Byzantine capital were assigned monthly shifts to treat patients here and in other hospitals of Constantinople.

Christian hospitals. From their beginnings, the Christian hospitals of Byzantine cities were meant for the poor, but as these institutions became increasingly advanced medical centers with the best physicians, some middle-class and even a few wealthy patients began to use them. In this regard Byzantine hospitals were different from the medieval West, where the middle class and nobility viewed hospitals as institutions meant only for the poor.

Islamic hospitals. Medieval Islamic hospitals were as good as the Byzantine medical centers. The first Islamic hospitals were founded in Baghdad during the reign of Harun al-Rashid. According to a governor of that time, Islamic hospitals had become commonplace by 820. Muslims considered support of hospitals a mark of true piety.

Although Islamic hospitals developed later than Christian institutions, they went through a unique development. They were very different from the Byzantine hospitals because they included separate sections for mental patients. Gradually these psychiatric wards became the most outstanding features of Islamic hospitals. Neither Byzantine nor medieval Western hospitals had wards for mental patients.

Early European hospitals. Hospitals developed more slowly in the western Roman Empire. Saint Jerome mentioned two small hospitals near Rome about 400. During the early Middle Ages, social conditions slowed hospital development in western Europe. Barbarian invasions from the north and Muslim advances from Africa discouraged a stable political, economic, and social life. Few

As was also true of Islamic institutions, Christian hospitals, such as the one in Siena depicted here, were supported by charity.

towns that were big enough to support important medical centers survived. In the territory of Charlemagne, hospitals did not develop beyond very simple hospices. As late as the thirteenth century, hospitals were rare in Europe. None of the 112 houses for the sick in medieval England provided physicians for their patients, nor did they store any medicines.

The Middle Ages

In the twelfth century, a new religious order, the Knights of the Hospital of Saint John of Jerusalem, known later as the Knights of Malta, reintroduced specialized medical care for the sick when they built their hospital in Jerusalem. The Knights served sick people in their hospital as vassals served their overlords. They became servants to their patients. As the Knights membership grew, they built many hospitals in the towns of Europe. In these hospitals they used rules they had set up in Jerusalem. Pope Innocent III established in 1200 the famous Hospital of the Holy Spirit in Rome, using the rules of the Knights hospital in Jerusalem as a model.

The Knights of Saint John had such a wide-ranging effect not only because their rule inspired western Europeans to help the needy, especially the sick, but also because Latin Christianity was entering a new phase of city-based growth. Many people began to move from the country to live in towns. These newcomers were exposed to crowded living conditions and to a wide range of diseases. Hospitals became necessary to treat the rapidly growing number of sick people in the cities and towns.

Inspired by the Jerusalem hospital, people in the area of Tuscany, Italy, began building hospitals during the thirteenth century.

▶ A late 1870s photograph of the Hospital of Saint John of Jerusalem, a medical institution founded by the Knights Hospitalers, or Knights of Saint John, in the twelfth century.

A sickroom from the fifteenth-century Hotel-Dieu hospital in Beaune, France, is preserved as a museum. The curtained beds often held more than one patient, who were nursed by nuns.

Before 1300, the town of Siena built an institution that kept on its staff a physician, a surgeon, and a pharmacist. In 1288 Folco Portinari founded the Hospital of Santa Maria Nuova in Florence. By the fifteenth century, this institution had developed into an important center for medical treatment. Hospitals in Renaissance Italy, in medieval Constantinople, and in Baghdad show that charitable institutions were not isolated from scientific medicine in medieval times.

urban of the town or city rather than the country

MODERN HISTORY During the early modern period (1650–1870), hospitals in Europe's **urban** centers were shelters for the poor and working classes. They operated as instruments of religious charity and social control with little help from the medical profession. Whether the patients were Catholic or Protestant, hospitalization continued to be an opportunity for physical comfort as well as moral uplifting. More and more, hospitals came under the control of nonreligious groups, including municipal governments, **fraternal** organizations, and private patrons.

fraternal characteristic of groups whose members share common beliefs or interests

At the same time, more positive ideas of keeping a person's good health and being cured of sickness suggested that illness was not unavoidable, sinful, and a human burden. Instead, sickness might be controlled and eliminated. By the middle of the nineteenth century, hospitals came to be seen as institutions for physical help and cure. They could be places of early rather than last resort, especially for military personnel and the labor force. These views meant planning for better health care to be given to more people.

A Hospital for All Social Classes

Thanks largely to progress in medical knowledge and technology, the practice of medicine in Western society was greatly advanced in

therapeutic healing or curing

electrocardiogram a graphic tracing made by a machine that records and measures the electrical impulses that cause the heart's contractions; the device is used to diagnose heart disease

the early twentieth century. By 1900, upper- and middle-class patients in Europe and the United States were seeking out medical care in hospitals. Staffed by competent medical professionals and equipped with medical laboratories and other tools for diagnosis (determining disease or health states), hospitals became the preferred place for surgical and medical care. Women began using hospitals for help in giving birth to and caring for new babies.

Advanced diagnostic and **therapeutic** procedures offered in hospitals after 1900 transformed hopeful ideas of physical rehabilitation and cure into reality. The use of X rays, **electrocardiograms**, and the medical laboratory greatly improved the ability of hospital personnel to make more accurate diagnoses. In addition to providing rest and a healthier diet, hospitals focused more and more on managing severe illnesses, especially life-threatening conditions that required intense, highly technical care.

The changing status of nurses and physicians. For patient care, hospitals began to use better-trained nurses, who were educated in professional programs based on the model established by Florence Nightingale (1820–1910). These new hospital nurses gradually replaced the decreasing number of religious staff members who had performed patient services. The Nightingale nurses became valuable assistants to the medical profession in patient management.

The scientific hospital. By the 1910s, more physicians joined hospital staffs, linking their professional reputations to the achievements of scientific medicine that hospitals seemed to make possible. In voluntary hospitals in the United States, medical staff organizations were flexible, allowing admission to both local general practitioners and specialists who could bring into the hospitals patients able to pay for their treatment. In Great Britain there were barriers between general practitioners, on the one hand, and

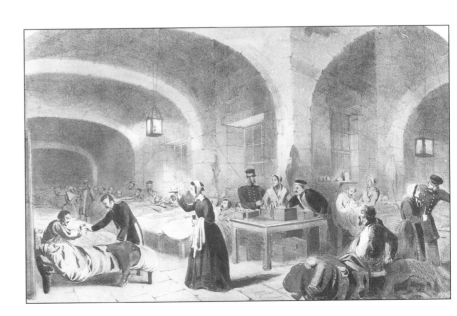

Florence Nightingale ministers to British patients lying on cots in a crowded sickroom in a Turkish hospital during the Crimean War (1853–1856).

▶ Entrance to a modern hospital.

hospital-appointed physicians and surgeons, on the other. Although they were encouraged to send their patients to hospitals, general practitioners were not allowed to practice in these hospitals.

The Hospital as Biomedical Showcase

Following World War II, the hospital rapidly strengthened its role as the provider of excellent scientific and technologically advanced medicine. Tremendous growth in medical knowledge led to the growth of diagnostic and therapeutic services at hospitals. This development had far-reaching implications for patient access to hospitals, hospital costs, and the quality of care provided to a wide range of people under various private and state-sponsored health plans.

The role of government. In the United States, the federal government's involvement in sponsoring hospital care gradually increased as the demand for hospital beds and services grew. Beginning with the Hill Burton Act in 1946, federal authorities supported the existing system of private hospitals. This supportive role helped to save a network of independent and competing municipal, religious, and university hospitals.

Physicians in hospitals. To work in hospitals of their choice, all practicing physicians in the United States must acquire admission "privileges" in such institutions. Most hospital care is provided by private doctors who visit the hospital to check on their patients and confer with hospital staff. This system allows for large and mobile medical staffs whose authority in the hospital is general.

Conclusion

The growth and change of the hospital in recent centuries make people wonder whether care is still the primary purpose of this institution. The hospital's original purpose was to shelter and comfort

Large hospitals are often the only new buildings in poor urban areas. They may be surrounded by boarded-up housing and parking lots where apartments used to be. There is debate over how much of its resources a society should devote to high-tech hospitals when the need for housing and schools and parks is just as great.

In the United States since 1980, many charitable hospitals have been converted to for-profit hospitals. Some hospitals have closed their emergency rooms, so that they do not have to provide emergency care to people who cannot pay for this service. The mission of the original hospital as a place for poor people has been completely lost.

all suffering people who needed medical help. To a great extent, hospitals now restrict admission to seriously ill patients who require the most advanced diagnostic and therapeutic measures. This development, combined with the high cost of treatment in a hospital, has made hospital stays rare and brief for most people. Three centuries of medical care have transformed the hospital from a caring shelter for the poor into a disease-oriented organization for the sick who can afford to be cured.

> Modern hospitals are not set up to treat people with chronic illness or people whose lives cannot be saved. Patients who are dying are better off in a hospice. A hospice does not treat patients but makes them as comfortable as possible.

"What is the moral responsibility of a hospital? Do you assure basic services for the many, or do you retain complex, sophisticated services that tend to affect the relatively few?"

—Robert D. Grumbs, Director,
New York City Health
Systems Agency (1989)

Hospitals have to deal with many ethical concerns and problems. All groups involved in some way with hospitals—patients, physicians, employees, board members, volunteers, the community at large, payers, business partners—have a stake in the way these ethical issues are considered and resolved.

Therapies: Hospice and End-of-Life Care

Related Literature

Many poets have given vivid images of their experiences in hospitals. A collection of poems by different poets, *Articulations: The Body and Illness in Poetry*, edited by John Mukand (1994), devotes a whole section to poems on the medical environment, especially the hospital, where patients come to grips with pain, fear, and embarrassment, often with a sense of estrangement and isolation. Several poets speak of the scary effect of technology: the X-ray machines, heart monitors, dialysis machines, and ventilators.

MEDICAL CODES AND OATHS

This entry consists of two articles explaining various aspects of this topic:

History
Ethical Analysis

HISTORY Ethical codes and standards are usually written by someone who is a member of the group to which the standards apply. Ethical standards for doctors are usually written by doctors, and standards for nurses are usually written by nurses.

Ethical standards may also be written by members of religious groups, cultural groups, or national and international organizations. Usually the standards are written as a "code of ethics" stating what people should and should not do. They can be written as prayers, oaths, rules, or statements.

An ethical statement or standard is a clear description of what it is morally right to do or not do. Each member of the group is responsible for following the standard. Groups of health-care providers may watch to make sure their members follow the standard. Government may also enforce the standards.

Documents Created by Practitioners

Throughout history, health-care givers have created prayers that give thanks for divine blessings and that ask for divine help in healing.

Medical prayers. In ancient Greece, the god of healing was called Asclepius. Healers prayed to him and asked that they be "like God: savior of slaves, the poor, of princes, and a brother to all." They promised that they would help everyone equally, no matter how rich or poor.

Ancient Jewish texts contain prayers for doctors to use. One is called the Daily Prayer. It asks God for courage, determination, and inspiration to help the physician develop skills, do what is necessary, and heal patients. The physician promises to put the patients' needs over his own, and prays that his healing is God's will for the patient.

Oath for physicians: *Charaka Samhita.* In ancient times, physicians often showed their desire to do the right thing by taking oaths. The oaths were promises, an important part of the initiation ceremony for medical apprentices. People believed that in order to be a good healer, the healer had to "work with" the gods in the treatment of disease.

One of the oldest known oaths is a medical student's oath from the *Charaka Samhita*, an ancient Indian manuscript. This manuscript contains concepts that have been an important part of Indian medicine since long before the birth of Christ. The medical student pledges to live like a slave to the orders of his teacher, with no luxuries. The student promises to put the patient's needs above his own, serve day and night with heart and soul, avoid drunkenness, crime, and adultery, and keep the secrets of healing that his teacher has passed on to him.

The Indian oath is different from oaths in the Western world, which require that the healer serve everyone equally. The Indian oath states that the healer must not give help to enemies of his ruler, evildoers, women who are alone, and people who are about to die! Even though there are differences in the oaths, they are similar enough that medical ethics may have spread from India to the West.

Hippocrates (3rd or 2nd century B.C.E.).

Medieval illustration of the Hippocratic Oath. Until recently, it was administered to graduates of most medical schools in the United States.

Oath of Hippocrates. The most lasting medical oath of Western civilization is the Oath of Hippocrates. Although it is well known, no one knows exactly when it began or who wrote it. It is part of the Hippocratic Collection of medical writings, which was organized and edited sometime after the year 400 B.C.E. by librarians in Alexandria, Egypt.

The Oath of Hippocrates has two parts. The first is a contract between student and teacher, and the second is an ethical code. In the opening sentences, the new physician, who in ancient times was always male, promises to become an adopted member of his teacher's family, to help support his teacher and his teacher's children if they need it, and to teach his instructor's children without pay. The student also promises never to share his medical knowledge with anyone who has not made the same promises.

The ethical code in the Oath of Hippocrates controls what the physician can and cannot do, and explains his relations with the patient's family. The physician who takes the oath agrees to make sure that patients get healthy food, and to protect them from harm and injustice. He promises not to tell the patient's secrets. The oath ends by asking that people give the physician who keeps the oath fame "for all time to come." If the physician breaks the oath, he will be looked down on "for all time to come."

For centuries after the appearance of the Hippocratic oath, physicians did not show any signs of accepting it or following it. Ancient Greek physicians ignored its rules. Some people believe that the rise of Christianity, which has ethical ideas similar to the oath, helped to spread acceptance of the oath. But there is very little evidence that early Christians had any interest in it. In fact, there is more evidence that there were important ethical differences between Hippocratic tradition and Christian tradition.

Revival of Hippocratic oath in the Middle Ages. No one knows why the oath became widely used in Europe in the Middle Ages. Some people believe that Europeans saw similarities between Christian and Hippocratic ideas. As people became more interested in the oath, they rewrote parts of it so that it was more in harmony with Christian beliefs and practices.

In modern times, the oath has continued to be used as a model for ethical pledges by physicians. Whether this is good or bad is open to interpretation. On the one hand, the oath is against exploitation of the sick, who are often the most vulnerable people in society. On the other hand, it assumes that the physician is the sole authority about what is good or bad for the patient. This authority conflicts with the late-twentieth-century idea that patients have a right to decide for themselves what they need and want in medical care.

In the eighteenth and nineteenth centuries, Western medical schools had their graduates take oaths. This was thought to impart high ethical standards to the student. It is unclear how often the

Hippocratic oath was used, but it is known that the typical oaths included Hippocratic ideas.

Declaration of Geneva. The Declaration of Geneva was written in 1948 by the newly organized World Medical Association. This oath is similar to the Hippocratic oath, and most U.S. medical schools use it. The declaration tries to make the original oath apply to modern medical practice and to different cultural, religious, and ethnic groups in modern society. This brings up the serious question of how any one ethical code could apply to many different religious and cultural groups, who may have very different ideas about what is or is not ethical behavior. The Declaration of Geneva does not have any references to God, or to religious beliefs. It may appeal to nonreligious physicians, but it also may offend physicians who are religious.

The United Nations building in Geneva, Switzerland, where the Declaration of Geneva was signed in 1948.

Declaration of Geneva

The Declaration of Geneva (1948) is an updated version of the Oath of Hippocrates, which is now almost 2,400 years old. The Declaration of Geneva reads:

Now being admitted to the profession of medicine I solemnly pledge to consecrate my life to the service of humanity. I will give respect and gratitude to my deserving teachers.

I will practice medicine with conscience and dignity. The health and life of my patient will be my first consideration. I will hold in confidence all that my patient confides in me.

I will maintain the honor and the noble traditions of the medical profession. I will not permit considerations of race, nationality, party politics or social standing to intervene between my duty and my patient. I will maintain the utmost respect for human life. Even under threat I will not use my knowledge contrary to the laws of humanity. These promises I make freely and upon my honor.

Although the writers of the Declaration of Geneva claimed that it simply updates the Hippocratic oath, there are major differences between the two. The Hippocratic promise to keep the patient's most serious secrets is replaced by a much tighter one to keep all of the patient's secrets, no matter what. This stricter promise may cause conflict in some situations, where the doctor may have to tell the patient's secrets in order to protect someone else from harm.

The physician who makes the declaration promises not to let the patient's religion, nationality, race, political party, or status interfere with his or her duty. The writers of the declaration all agreed to make this very clear, because they felt that the Hippocratic oath was not explicit enough.

The Declaration of Geneva has been accepted by some medical professional groups, but national organizations of physicians have not adopted it. In addition, it conflicts with the ethical ideas of many religious and nonreligious people both in the East and the West.

Taoist code. One of the earliest codes of medical ethics comes from China, where physicians were not familiar with the Hippocratic oath. Beginning in the seventh century, a Chinese tradition of medical ethics developed. The Taoist philosopher Sun Szu-miao stressed the importance of preserving life and of putting compassion for the patient over the healer's self-interest. His work shows that in his time there was an elite group of "great physicians," which marks the first time a group of physicians claimed they had special medical authority.

British codes. In the West, the Royal College of Physicians in England had an interesting example of a professional code. In the first rules of the college, written in 1555, and in the revision of 1647, there is a section called "Moral laws and their penalties." This section requires good behavior in college meetings, regular attendance, and proper **etiquette** between several physicians who are working with the same patient.

In 1803, Thomas Percival (1740–1804) of Manchester, England, wrote a code of medical ethics. It advises physicians to treat patients with "tenderness, steadiness, condescension, and authority." This description shows the attitude of a wealthy English gentleman kindly giving benefits to patients, who are expected to show proper gratitude. Percival's Medical Ethics is in the Hippocratic tradition but is a beginning in acknowledging that physicians have a debt to society as well as to patients. Unlike the Hippocratic oath, which makes physicians swear they will not do any kind of surgery or prescribe drugs, Percival's code makes surgery and the prescription of drugs acceptable.

Percival wrote his code to help solve a dispute between surgeons, physicians, and apothecaries (people like pharmacists, who prescribed and provided drugs). Because of this, his code has many comments on etiquette and relations between different healthcare providers. It has procedures for physicians to help each other in difficult cases and other procedures to keep everyone aware of who is the most important or highest-ranked physician in a hospital or in a group working with the same patient.

Percival believed that criticism of physicians is usually not based on any real problem, and that it was also degrading to the doctors who were criticized and to the profession of medicine in general. His Medical Ethics calls for doctors to act in a way that makes the public have more respect for them and for the entire medical profession.

First American code. Some of the earliest American work on physician ethics was written by Samuel Bard, who was a doctor at Columbia University in the late eighteenth century, and Benjamin Rush, a Revolutionary War patriot. The medical associations of Boston

see also

Religious Perspectives: Taoism

etiquette rules of conduct in social or professional relations

▲ Benjamin Rush, American patriot and physician, wrote the very first American work on medical ethics.

The American Psychiatric Association has issued detailed Guidelines on Confidentiality with regard to special situations, records, and the legal process. The American Bar Association has offered a handbook that addresses the issues of confidentiality, consent to disclosures, third-party access to information, and penalties for unauthorized disclosures. The Council on Ethical and Judicial Affairs of the American Medical Association outlined in 1992 the limits and value of confidentiality. All these documents stress individual rights and specify professional responsibilities regarding confidentiality.

and Baltimore and the state of New York also wrote codes. When the American Medical Association (AMA) was organized in 1847, it used Percival's Medical Ethics and other documents as the basis of its code of ethics. This code of ethics did not mention etiquette for doctors working in hospitals and barely referred to the relations doctors had with pharmacists and courts of law. It expanded and gave more details on physicians in private practice, and even gave a statement on the obligations that patients and the public had to doctors.

International codes. In 1948, the World Medical Association was established. The association encouraged physicians to develop international standards of medical ethics. The organization adopted an International Code of Medical Ethics in 1949. This code attempted to summarize the most important principles of medical ethics.

Since 1900, the number of unqualified people practicing medicine diminished thanks to laws that made it necessary for doctors to be certified before they could practice. Trained physicians were better at their work because of scientific advances. By the middle of the twentieth century, physicians were paying more attention to the treatment of patients and less to the etiquette of relations between doctors, or between doctor and patient. The International Code shows these changes. Instead of detailed descriptions of etiquette, it simply states that doctors should behave toward other doctors as they would like other doctors to behave toward them. It also recommends that they call in specialists in difficult cases, and that they should not try to take patients away from other doctors.

The International Code only hints at the ethical problems of abortion and mercy killing by mentioning the doctor's responsibility to preserve life. It states that doctors should avoid doing anything that will weaken the patient but will not make the patient any better. This seems to be a mild warning against mercy killing, which would involve weakening the patient to the point of death, without any intention to cure.

Principles of Medical Ethics. In 1957, using the International Code of Ethics as an example, the AMA reduced its lengthy code to ten one-sentence Principles of Medical Ethics. This was intended to clarify its code, not shorten it, although it had that effect. The association kept the Hippocratic idea that the patient's benefit comes above everything else, but added that the doctor is also responsible to society, not just to the patient.

Most of these ten principles are also in the International Code of Medical Ethics, but a few are not. The 1957 principles warn doctors against associating with other health-care providers whose methods might be unscientific.

In the late 1970s, the principles were revised again. A special committee was formed to clarify and update the language. The committee also eliminated references to doctors as "he," since many doctors were now female, and balanced ethical standards with current legal ones.

For the first time, the principles mentioned patients' rights: "A physician shall respect the rights of patients, of colleagues, and of other health professionals." This was a change from the traditional Hippocratic oath, which gives the doctor complete authority to decide what is right for the patient.

Codes from Outside the Profession

A fundamental question of ethics underlies the development of laws and codes governing medical conduct: Are doctors as a group responsible for determining what is proper and right conduct? Or should the public decide this? Even if it is decided that doctors should determine the ethical code for their profession, it is not clear which doctors should have the authority to determine what the codes should be and speak for their entire profession.

In modern times, medical ethics has often assumed that doctors should define their own codes of conduct. This has not always been true. Religious groups and governments have sometimes claimed the right to decide on ethical issues. More often, professional groups and the public insist that ethical judgments should not be made only by members of the professions. Many people believe that ethical conduct should be based on broad cultural, philosophical, and religious ideas.

Governmental codes. In the twentieth century, governments of many nations used ethical codes as a basis for laws governing the medical profession. These laws were enforced by an official, publicly appointed medical board. These laws sometimes parallel the principles of the Percival tradition, but many of them deal with problems that are unique to modern society. Many of them also show the modern concern for both public and individual welfare.

The Nuremberg Code. Some of these codes deal with single subjects. For example, the Nuremberg Code, which is an international set of laws, regulates medical research on human subjects. This code was written as a result of World War II, following public outrage over Nazi experiments on people in concentration camps during the war.

The oath of Soviet physicians. In what was then the Soviet Union, a government-sponsored medical oath was adopted in 1971, when the Presidium approved the Oath of Soviet Physicians. This oath was modeled after one that had been used at the University of Moscow since 1961. The physician pledged to obey Communist principles and vowed responsibility to the Soviet government. No other medical oath had ever included this commitment to specific political beliefs and national government.

The Soviet oath did not leave out other moral obligations, though. It instructed the physician to honor professional secrets, constantly improve knowledge and skill, always be available to calls

The many historic codes and oaths of health-care professionals emphasize four principles:

1. nonmaleficence ("Do no harm")
2. beneficence ("Do good for patients")
3. confidentiality
4. justice

None of the codes and oaths refer to:

1. patient autonomy
2. truth-telling to patients

for medical care or advice, and dedicate all knowledge and strength to professional activities. Like other late-twentieth-century oaths, the Soviet oath included the same ideal of humanitarian duty to individual patients that appears in the oldest medical oaths, but it also pledged the doctor to serve the interests of society.

Nongovernmental groups. Throughout history, ethical codes, prayers, and oaths have also been sponsored by private groups, religious groups, and consumer groups that do not represent the medical profession.

For centuries, the Roman Catholic Church has provided moral views about medical matters. These matters include **abortion**, mercy killing, and birth control. Since medieval times, the Church has presented these views in systematic theological writings, in descriptions of morally confusing cases, and in manuals of theology.

The Church's views on secrecy, patient consent, organ transplantation, and care for people who are terminally ill are very similar to those of other codes. The Church is against abortion, except in cases where it is an accidental result of another treatment that is intended to protect the mother. For example, if the mother has cancer and has to have chemotherapy, this may result in abortion of the fetus. The Church prohibits sterilization of both men and women, unless it is a treatment for a serious illness, and it prohibits artificial insemination. The Church's directives follow the Vatican's **encyclical** *Humanae Vitae*, "Instruction on Respect for Human Life" (1968).

abortion the deliberate or spontaneous early ending of a pregnancy, resulting in the death of a fetus

Religious Perspectives: Roman Catholicism

encyclical a letter from the pope to all the bishops of the church or to those in one country

Greek fresco showing Asclepius, the god of healing, treating a patient in a sickroom.

Conclusion

Professional leaders, patients, and public policymakers who are involved in establishing ethical standards on new issues are confronted with many difficulties. These difficulties reflect the conflicts in values among different people and groups, who have differing views on medical ethics. Traditional professional ethics of physicians emphasize the virtue of benevolence and the doctor's responsibility to serve the patient. This tradition honors the relationship between doctor and patient, professional secrecy, and the doctor's duty to work for the patient's good. Traditionally, physicians' ethics have given the doctor the main authority to determine what is good for both the doctor and the patient.

Codes written by nonphysicians, including those written by religious and government groups, have given other philosophical ideas as a basis for ethics. Some of these codes have emphasized justice or equality in treatment. The differences between these codes and the traditional ones have resulted in increased confusion, because doctors are subject to competing ethical codes, which have conflicting standards.

In the late twentieth century, responsibility for development of ethical codes began shifting from doctors to society as a whole. When there is no ethical guideline for a situation, the responsibility for deciding what is ethical is partly the doctor's, partly the patient's, and partly society's. The future success of ethical codes may depend on the combination of physicians' ethical views with those of society as a whole.

ETHICAL ANALYSIS Over the centuries, codes, oaths, and prayers of medical ethics have come from many different sources. It is not surprising that they are very different from each other, both in style and in content.

It is difficult to do an ethical analysis of codes of medical ethics, because these codes are not complete and systematic theories of medical ethics. However, modern codes are usually the result of a great deal of discussion, debate, and review. The codes reflect the basic ethical views of the groups that produced them.

Most codes, especially the ones written by doctors, have a basic principle or ethical obligation that helps the doctor decide what to do in ethically confusing situations. This is useful, because sometimes the codes contradict each other, and the ethical issues involved are often very controversial.

Hippocratic Oath

The ideas in the Hippocratic oath, which was written in ancient times, will likely last into the twenty-first century. The basic ethic of

---

In its February 25, 1998, issue, *The New England Journal of Medicine* published a doctor's contemporary and ironic understanding of the ancient Hippocratic oath. Here are excerpts from Dr. Barry M. Manuel's mock oath, which he created out of frustration:

1. I swear by Apollo, the Physician, [and] by Asclepius . . . to keep according to my ability and judgement the following oath:

2. To consider dear to me as my parents those teachers who taught me this art but did not inform me of the terrible climate in which I would have to apply this skill . . . Despite this, I shall look after my teacher's children as my own children, but will thoroughly acquaint them with realities of practice as it is today.

3. I will prescribe regimens for the good of my patients according to my ability and judgements but never violate the rules of Medicare, Medicaid, HMOs, and the like . . . I will always use generic drugs, even if I think that the proprietary drug is more effective.

4. As required by law, I shall share with federal and state agencies, insurance companies, professional review organizations, and the like, all strictly confidential information that has come to my knowledge in the exercise of my profession.

the Hippocratic oath is the doctor's promise to do what he thinks is best for the patient. This is repeated twice in the oath. Once it is applied to diet—the doctor promises to be sure the patient eats healthy food—and once it is applied to the doctor's good conduct when he visits the patient at home.

The Declaration of Geneva, written in 1948, repeats the idea that the doctor's first responsibility is to do what is good for the patient. The physician who takes this oath promises, "The health of my patient will be my first consideration." This idea is also included in the International Code of Medical Ethics of the World Medical Association, which says, "A physician shall owe his patients complete loyalty and all the resources of his science." The post-Communist Russian oath, the Solemn Oath of a Physician of Russia, has the physician pledge to work always for the patient's good.

The Hippocratic oath's individualism. The first characteristic of the Hippocratic oath is that it is mainly about the well-being of the individual patient. This is different from some other ethical systems. In utilitarian ethics, something that is good for a greater number of people, in the long run, is more ethically justified than something that benefits only one individual.

There is no evidence that the author of the Hippocratic oath, or the twentieth-century creators of oaths, believed in utilitarianism. Instead, they seem to believe that the doctor has a specific responsibility to benefit the patient, and that this responsibility is not related to any consequences for the society as a whole, or for others who are not patients.

Ethical problems happen in situations where the physician believes that an action will produce the most good for people in general, but a different action will be better to the patient being treated. A physician who chooses the action that benefits the patient most is being faithful to the oath and is rejecting the utilitarian ethic.

In 1957, the American Medical Association (AMA) wrote its Principles of Medical Ethics. It did not accept the Hippocratic emphasis on the individual. It instructs the physician that the main goal of the medical profession is to serve humanity as a whole.

paternalism the traditional idea that fathers know what is best for their children

The Hippocratic oath's paternalism. The central ethic of the Hippocratic oath is also paternalistic (like a father). This means that the physician, like a parent of young children, has the sole authority to decide what is best for the patient. The patient does not have a voice, and is not to decide what is best for himself or herself. In fact, the doctor pledges to protect the patient "from the mischief and injustice which he may inflict upon himself if his diet is not properly chosen."

This paternalism is also clear in the pledge in the Hippocratic oath that the doctor must keep medical knowledge secret and not tell medical secrets to people who are not doctors.

▶ A miniature illumination from the manuscript "Cirugia" (c. 1300) by Roger von Salerno shows a doctor treating patients with a variety of ailments, from stomach abscesses to broken ribs.

Thomas Percival, who lived from 1740 to 1804, believed that doctors should study not only tenderness and steadiness but also condescension and authority. He also believed that patients should be encouraged to feel gratitude, respect, and confidence in the doctor. The AMA principles of 1957 and the 1959 British Medical Association codes stated that the patient's medical secrets could be told to others if this was necessary for the patient's good.

Codes Written by Groups Outside the Medical Profession

Many of the codes that have been written more recently by governmental and religious groups have not shown these characteristics of individualism and paternalism. The Nuremberg Code was written in 1947 and was one of the first codes of medical ethics to become a part of international law. This code could have used Hippocratic principles to state that using human subjects for research is always wrong, because doing an experiment to gain general knowledge does not clearly benefit the person who is experimented on.

The Nuremberg Code did not use this general Hippocratic principle. Instead, it stated that sometimes physicians need to use human subjects in experiments in order to gain knowledge that benefits everyone in society. Physicians have a duty to ensure that subjects are willing and that they have a right to know all the risks of the experiment before consenting to be a part of it.

Consent

Other codes written by governmental and religious groups used the idea of patients' rights to show that they had broken away from the old medical oaths that focused only on consequences for the patient. This modern emphasis on rights is influenced by the authors of the Bill of Rights of the United States Constitution.

The focus on rights and duties emphasizes the patient's right to give informed and voluntary consent to being used for research and to any part of medical treatment. The idea of consent is completely missing from the old, traditional codes written by medical professional groups. Consent is based on the ethic that each person has the right to determine what is best for him or her, and it is also based on the legal idea that each person has a right to choose what he or she wants to do.

This introduction of rights and duties, and the underlying moral idea that each person deserves respect, shows a rejection of traditional Hippocratic paternalism and consequentialism. It also provides a way of moving away from pure individualism, in which everyone does whatever he or she wants to do without thinking of the consequences for other people.

> The first health-care association that used the language of rights was the International Council of Nurses' Code for Nurses, which was written in 1973 and reaffirmed in 1989.

Specific Ethical Injunctions

The Hippocratic rule of keeping the patient's secrets is sometimes taken to forbid telling medical secrets. The text is actually rather vague. It says, "Whatever I may see or hear in the course of treatment in regard to the life of men, which on no account one must speak abroad, I will keep to myself, holding such things shameful to be spoken about." Each doctor must decide what is meant by things that "on no account must be spoken abroad." Possibly, doctors are expected to decide whether or not to reveal a secret based on whether or not this will benefit the patient.

Confidentiality in the British and international codes. In the British Medical Association's 1959 ethical code, confidentiality was the principle physicians were told to follow. The code said also that the complications of modern life sometimes make it difficult for the doctor to know when to apply the principle of confidentiality, and that sometimes the rule must be modified. How should the doctor decide when to break the rule? The standard to follow is always to do what will benefit the patient, or what will protect his or her interests.

The World Medical Association's International Code of Medical Ethics (written in 1949, and amended in 1968 and 1983) and the Declaration of Geneva (written in 1948, and amended in 1968 and 1983) both deny that there is any kind of excuse for telling the patient's secrets. They clearly require "absolute secrecy," like an ancient Jewish oath, the Oath of Asaph.

Exceptions to the confidentiality rule. The 1957 AMA Principles, which were revised in 1971, are now outdated, and so are the 1973 Principles of the American Psychiatric Association, which were based on them. Both of these sets of principles clearly give three exceptions to the rule of confidentiality. The doctor may tell the patient's secrets if he or she thinks this will benefit the patient, but also if the doctor thinks it will benefit society, or if it is required by law. For example, the doctor would tell the police if a patient had a bullet wound that the patient received while committing a crime.

The ethical problem in these broad exceptions is not just the fact that the doctor has the sole power to decide when it is beneficial to tell the patient's secrets. It is also a problem because the patient's right to privacy is considered secondary to society's interests.

Abortion. Abortion is a subject that many people have strong opinions about. Groups that have written ethical codes have usually followed their own ethical views about it. The author of the Hippocratic oath was a follower of the Greek philosopher Pythagoras, and the oath follows the Pythagorean belief that abortion is wrong, even though in the Greek culture of the time abortion was not considered unethical. In the early medieval Jewish Oath of Asaph, the medical student is told not to abort an unwed mother's baby—"Do not prepare any potion that may cause a woman who has conceived outside of marriage to miscarry."

Abortion and the Catholic Church. The 1975 Ethical and Religious Directives for Catholic Health Facilities follow the Church's strong belief about abortion. Of forty-three principles, seven relate to abortion. Ending a pregnancy on purpose, before the fetus is developed enough to live outside its mother, is not permitted. Neither is intentionally destroying a fetus that is able to live. If a treatment is intended for some other purpose, but unintentionally ends the pregnancy, it is allowed, but only in cases where the life of the mother is at stake and when the treatment cannot wait until after the fetus is developed enough to live outside the mother.

Abortion and international organizations. If the group that writes the ethical code represents a diverse culture with many different ethical ideas, the code is always less specific about whether or not abortion is ethical. In an early version of the World Medical Association's International Code, the association said that abortion may be performed only if the doctors performing it believed it was acceptable, and if it was not against the law in the country where it was done.

Abortion and health professionals. The American Nurses' Association also represents a wide variety of people, and it also avoids a direct decision about abortion. In its code, which was rewritten in 1968 and in 1978, the association says that "the nurse's

> "It is very little to me to have the right to vote, to own property, etc., if I may not keep my body, and its uses, in my absolute right."
>
> —Lucy Stone (1818–1893), founder, National American Woman Suffrage Association

> "What is this life by which you, who exist still incomplete, count for more than I, who exist complete already? What is this respect for you that removes respect for me? What is this right of yours to exist that takes no account of my right to exist?"
>
> —Oriana Fallaci (Italian journalist), *Letter to a Child Never Born*, 1975.

> "The greatest destroyer of peace is abortion because if a mother can kill her own child, what is left for me to kill you and you to kill me? There is nothing between."
>
> —Mother Teresa (1910–1997)

respect for the worth and dignity of the individual human being extends throughout the entire life cycle, from birth to death." Since this does not mention the time before birth, it seems as though this statement does not apply to fetuses.

Mercy killing. It would seem obvious that medical oaths would all make the doctor promise to preserve life. However, most do not mention this at all. The only code that does mention it is the World Medical Association's International Code, which was revised in 1968 and 1983. This code says that "a physician shall always bear in mind the obligation of preserving human life."

It is much more common for the codes to prohibit active killing. In many cultures, killing is forbidden, even if it is considered mercy killing. The Hippocratic oath states, "I will not give a deadly drug to anybody if I am asked for it, and I will not suggest the use of a deadly drug." This statement is controversial, and people interpret it in different ways. Some people say it refers only to murder. It seems more likely that it refers to mercy killing or helping a patient to commit suicide.

In the codes of professional groups, or codes written by the Roman Catholic Church, giving a deadly drug, or helping a patient to die, is not considered the same as withdrawing treatment from a terminally ill patient and allowing the patient to die naturally, even if the patient dies sooner than he or she would have if the treatment had continued.

The AMA states in its Judicial Council Opinions that "the physcian should not intentionally cause death." But it also recognizes that if the patient is on life-sustaining machines, without any hope of recovery, sometimes the patient wants the machines to be turned off, which leads to death. The post-Communist Russian oath follows the Hippocratic oath and forbids the doctor to give a deadly drug.

The distinction between active killing and mercy killing is made clearer in the American Hospital Association's document, A Patient's Bill of Rights. This document says that the patient has the right to refuse treatment "to the extent permitted by law," even if the result is the death of the patient. However, the patient does not have the right to drugs that will actively cause death.

Truth-telling. One conflict between the principle of "always do what's in the patient's best interest" and the ethical theories that are based more on the professional's duty to the patient is over the question of what doctors should tell a patient who is dying. The Indian oath of the *Charaka Samhita* is very clear. It says that the doctor should not tell the patient that he or she is dying if this news would be a shock to the patient or to others. The 1847 version of the AMA code states that a doctor should not tell the patient that he or she is likely to die, unless it is definite. Then the doctor should tell the patient's friends and family, and then—if it's absolutely necessary—the

> "Prolongation of life should not be the aim of medicine. Health is the aim of medicine."
>
> —George F. Will,
> political columnist

patient. These codes do not recognize that telling the patient's family and friends first is a violation of the patient's confidentiality. This violation is justified by saying that it is best for the patient.

Even after 1980, when the AMA principles promised complete honesty to the patient, the view persisted that the doctor knows what is best for the patient, and can keep information from the patient, even if the information is that the patient is going to die. The association's Council on Ethical and Judicial Affairs stated that the requirement that the patient be given enough information to make informed choices about treatment can be broken. If telling a patient that he or she may die is seriously upsetting the patient, the Council decided, the doctor does not have to tell.

Justice in delivering health care. Many ethical codes of doctors, and other medical care ethics, talk about equal treatment in health care. The Hippocratic oath uses the Greek word *adiki'e*. This is often translated into English as "justice," but it does not refer to equal treatment or equal distribution of benefits.

Equal access to medical treatment is not recognized in the Hippocratic oath, but it is an ideal in many modern codes. The twentieth-century Declaration of Geneva states: "I will not permit considerations of religion, nationality, race, party politics, or social standing to intervene between my duty and my patient." The American Nurses' Association code says, "The nurse provides services with respect for the dignity of man, unrestricted by considerations of nationality, race, creed, color, or status." The AMA recognizes that society must decide how health-care resources, which are limited, are divided up among people. It urges that health care be provided based on "fair, socially acceptable, and humane criteria." It also emphasizes that the doctor's duty is to do everything possible to benefit the individual patient. The post-Communist Russian oath pledges never to deny medical assistance to anyone and to provide equal care to patients regardless of their economic status, nationality, or religion.

The Ethics of Professional Relations

In contrast with public codes or bills of rights, almost all codes written by health professionals pay a lot of attention to relationships between health-care providers. The Hippocratic oath begins with the physician's promise that he will treat his teacher like a parent, that he will give his teacher money or other aid if necessary, and that he will treat his teacher's children like brothers and will teach them for free. It also includes a promise to keep the medical secrets his teacher has given him.

How do doctors relate to each other? At least two different ethical views might be the basis of the detailed descriptions of obligations that medical workers owe each other. First, these duties could be seen as courtesies that all people owe each other, whether they are doctors or not. For example, the AMA code of 1847 states

> "If criminals have a right to a lawyer, I think working Americans should have the right to a doctor."
>
> —Harris Wofford,
> United States Senator,
> *The Washington Post*, 1991

▶ In the eighteenth century, it seemed necessary to regulate relations between doctors and between doctors and patients.

▲ A substance-abuse counseling ad in the classified section of a newspaper.

that if a physician is qualified, he cannot be excluded from working with other doctors, and that when doctors work together on a case, they should arrive on time. They should also agree to respect the patient's privacy, and if they disagree about what treatment is best, they should not let the patient hear them arguing.

A second possible reason for the obligations medical professionals have to each other is that professional duties are in the interest of society. In other words, these duties benefit everyone, but only members of the profession are trained enough to see this. For example, doctors historically have not advertised their services. They said that anyone who advertised was obviously just a salesman of medical care, and was not ethical. So in a way advertising, or the lack of it, was supposed to protect society at large by letting people see the difference between ethical doctors who did not advertise and the unethical ones who advertised.

If doctors believe they have special ethical obligations, but these obligations do not make sense to people who are not doctors, then there are likely to be conflicts between what doctors think they are ethically obliged to do and what society in general thinks is right. Even if doctors agree that they have a duty to preserve life or to limit advertising, the public at large might not agree with these views and may not want doctors to follow them at all times.

Advertising. A specific example of this involves the ethics of advertising. Many professional codes, such as the 1847 code of the AMA, prohibit or restrict advertising by members of the profession. This restriction on advertising might be seen as an agreement by members of the group to restrain price competition among themselves. Or, doctors may really believe that they provide a moral

General topics: Advertising

service that should not be sold as if it were a car wash or a dozen eggs. Whether or not doctors believe advertising is unethical, the public may believe it is unethical not to advertise, simply because it restricts competition among doctors and keeps prices high.

Conclusion

Ethical codes, oaths, prayers, and bills of rights come from different sources. They represent different professional groups, public agencies, and private nonprofessional organizations such as churches and patients' groups. It is not surprising that each group has different opinions about what is ethical or not, and that ethical codes are based on different ethical theories and methods of reasoning.

A big problem health professionals and the public face is how to decide what ethical codes should be used when a person is a member of more than one group. A person may be a member of a professional group, but at the same time may be a member of a religious or cultural group that has very different moral views. For example, the American Nurses' Association's code might be interpreted as allowing a woman to choose abortion. If a nurse is a member of the Roman Catholic Church, which prohibits abortion, what is he or she to do when a patient wants an abortion? Or, what if a non-Catholic nurse, who believes a woman has the right to choose an abortion or birth control, works in a Catholic hospital, where these choices are prohibited?

Conflicts for people who are members of more than one group come up for many health-care professionals. Also, people may reach their own decisions about what is ethically right, and their decisions may not match any code, whether it is written by a professional group or by religious, cultural, or governmental groups. In order to come up with an answer for these questions, each individual must understand the ethical differences and implications of all the codes.

Related Literature

Elizabeth Bishop's poem "The Staff of Aesculapius" (1956) somewhat resembles the shape of the snake-entwined staff that has become the symbol for medicine. She describes the staff as a symbol of Hippocrates, who was the first to substitute skill and experiments for superstition in treating sick people.

NURSING

This entry consists of two articles explaining various aspects of this topic:

▲ Florence Nightingale.

NURSING AS A PROFESSION Care for the ill or injured has existed since the beginning of recorded history, but modern nursing began in the nineteenth century with Florence Nightingale (1820–1910). Before Nightingale, there were no professional nurses. When people were sick, they relied on family members or friends for care.

Nightingale believed that people should care for the ill or injured because it was a moral practice, simply the right thing for good people to do. She believed that nursing should focus on good hygiene and caring and should be an independent profession, not necessarily connected with established medicine and hospitals. She believed that nurses could work on their own, instead of working under the supervision of doctors.

Nightingale established the first school of nursing at St. Thomas Hospital in London. There, in spite of her beliefs, nurses soon became subordinate to doctors. This did not match Nightingale's original vision of nursing as an independent discipline.

Nursing was not thought of as a "profession" until nurses developed their own schools, increased their skills and knowledge, and worked to gain respect. In their effort to be recognized as a profession, nurses began to apply science to nursing in the 1960s. They began doing research on the best ways to care for patients.

Nightingale's Vision

"A new art and a new science has been created since and within the last forty years. And with it a new profession—so they say; we say, *calling*," Florence Nightingale wrote in 1893 to the meeting of the International Congress of Charities, Correction and Philanthropy in Chicago. This meeting began the organization of the nursing profession in the United States and Canada.

Nursing as a calling. Nightingale thought of nursing as a "calling" instead of a "profession," because she was a religious woman and believed that nurses were called by God to care for others. She wanted to make it clear that to her, nursing was a moral and religious act, not simply a medical one. In fact, she emphasized caring over science. She believed that good nursing care fulfilled the nurse's moral obligation to care for others by putting "the patient in the best condition for nature to act upon him." In Nightingale's time, people did not fully understand what caused illness. She

In the early 1900s, most nurses were young and single women, who provided an inexpensive source of labor for hospitals.

emphasized the fact that if you keep patients clean, well-fed, and well-rested, they will often recover naturally.

Nightingale's emphasis on morals was important at the time, because in the late nineteenth century, people thought women who worked outside the home were not moral. Because nurses worked outside the home, and because they cared for men who were not in their immediate families, the public impression of them was that they were prostitutes or drunks. Nightingale fought against this negative view. She saw nursing as a way for women to make positive contributions to society, and recruited only women she believed had high moral character.

Nurses lose their independence. As medical science advanced, nurses lost much of their independence from doctors and hospitals. Doctors were in charge of patient care, and nurses were seen as their helpers. Nightingale had been afraid this would happen. When people began to understand that illness was caused by germs, Nightingale worried that this scientific knowledge would change the emphasis of medical treatment from caring in its moral sense to purely scientific intervention, and it did. This new concept of patient treatment was called intervention medicine.

Intervention medicine involved the belief that doctors cured by stopping the development of disease, while nurses simply took care of those being cured, which was thought of as less important. Also, people believed that scientific activities were just for men. Caring was believed to be a female activity. Because women and women's work were looked down on, so was nursing.

The primary focus of caring for most women was their own family. Because of this, in the early part of the twentieth century, most nurses were young, single women who were planning to go on from nursing to fulfill what people thought was their real calling— raising and taking care of their own families. While they were students, these young women were a cheap source of labor for hospitals. These young women gave most of the direct nursing care in hospitals, directed by the few older women who had made nursing their permanent career.

Nursing since World War II. During World War II large numbers of women entered the work force for the first time, because so many men were away fighting. They took the men's jobs in such places as war factories, farms, and many clerical jobs. Nurses serving in the armed forces showed how important nursing was. But after the war, many people still believed it was acceptable for women to work outside the home only when war required it. During the 1950s, nursing did not seem to progress as a profession. Although married women were now allowed to enter schools of nursing and work in hospitals, traditional attitudes toward women and the value of their work had not changed.

Nursing as a woman's job. Throughout history, some men have provided nursing care for the ill and wounded. Generally these

men have been in the military or in religious service. Since the development of modern nursing, few men have chosen nursing as a career. Even though men were encouraged to become nurses in the late twentieth century, only 3 percent of nurses were men.

Throughout the world, nursing is still thought of as a woman's job. Some people believe that men do not become nurses because nursing is not an attractive career. The career structure is poor. Nurses do not advance much within the profession. There is often a lack of resources for nursing care. In many countries, hospitals have no budget for nursing. Men, who have more career options than women, may simply choose careers that do not have these problems.

Gaining Recognition as a Profession

Since the 1960s, nursing organizations and scholars have debated about whether nursing is really a profession. Researchers have listed standards of a profession, which are helpful in understanding whether nursing is a profession or not.

A long educational process. In the United States, nurses are trained in several ways: through diploma programs in hospitals, through two-year programs in community colleges, and through four-year or longer programs in colleges and universities.

In the latter half of the twentieth century, researchers began recommending that all nurses be college-educated. As early as 1965, the American Nurses' Association recommended that anyone who is licensed to practice nursing should have at least a four-year college degree. Although many hospitals have continued to educate nurses, more nurses are preparing for their careers through associate (two-year) and baccalaureate (four-year) college degrees.

American army nurses in 1942.

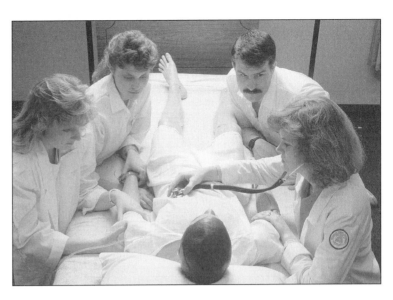

Nursing students practice on a dummy in a nursing lab at Brigham Young University, 1988.

Educational degrees in nursing. In the 1960s, nursing education began to expand to graduate education. Master's-degree programs developed in some nursing specialties, such as adult health, maternal and child health, and psychiatric-mental health. Doctoral programs in nursing developed more slowly, but between 1974 and 1984 the number of programs doubled, from twelve to twenty-seven. By 1993, the number had doubled again.

Authority and judgment. In the United States, since the beginning of the twentieth century, state boards of nursing have regulated nursing practice and made sure that standards for practice and education are met. The members of these boards are all professional nurses, and they make sure that nurses follow the standards the boards set and that they work safely and effectively. In the 1960s, the state boards created tests that nurses had to pass in order to be licensed to nurse. Without a license, a nurse cannot work.

The National League for Nursing (NLN), an organization of nurses and citizens who want to improve nursing, has developed standards for educational programs. The NLN has developed methods to check the quality of nursing education and to see whether a school meets these standards. If a school passes their NLN tests, it is approved. This approval is called accreditation.

The right to diagnose and prescribe. Nurses are licensed to practice nursing by state boards of nursing and state boards of medicine. Many states have passed laws allowing nurses to write prescriptions. Before this, only doctors could write prescriptions. Some states allow nurse practitioners and nurses in advanced specialties to receive payment from government programs like Medicaid and Medicare, and from private insurance companies. The amount of money a nurse receives for caring for a patient depends on the patient's diagnosis, or category of disease, and the treatment needed.

Nurses have often been taken for granted. A majority of nurses leave the profession after five years of work experience. The main causes of this trend are:

- underpayment
- overwork
- disrespect from doctors

Categories of illness and treatment needs were developed for doctors to use when they report what care they have provided for a patient. These categories are not useful for nurses, who provide different care, so the North American Nursing Diagnosis Association was created in the 1980s. This association developed categories of illness and treatment for nurses. Nurses use these categories to make sure they are providing the correct care for a patient, and to report the care to the agency that pays them.

Active and cohesive professional organization. The International Council of Nurses was established in 1899. It is an independent federation of nursing associations from many countries, but it is not connected with any national government.

In the United States, nursing has been moving toward primary care since the development of nurse practitioners. Primary care means that nurses act independently from doctors and are responsible for their actions. To achieve these gains, nursing has needed political power. Nurses gained political power in 1991, when the American Nurses' Association, the National League for Nursing, and the American Association of Colleges of Nursing joined to form the Tri-Council for Nursing Executives. The Tri-Council developed "Nursing's Agenda for Health Care Reform" in 1993. This agenda was supported by sixty-four nursing organizations.

The Tri-Council has led the effort to get the United States Congress to increase primary health care in communities. The Tri-Council also wants Congress to help communities to lead their citizens to take responsibility for personal health, live healthy lives, and learn about treatment so that they are fully informed about their options when they have to make health-care choices. The Tri-Council also encourages Congress to cut spending and use the health-care providers who give the best care for the lowest cost.

Strong level of commitment. A survey in 1981 reported that people thought nurses did important work, but also that people thought nurses were not very committed to their jobs, because only 40 percent of licensed registered nurses worked full-time. This seems wrong, since commitment cannot be measured by using only the percentage of nurses who work full-time. Commitment to nursing means the nurse's dedication to nourish patients' health and well-being. A study in 1984 found that nurses who were considered excellent workers were very committed to the patients' well-being.

Nursing journals, articles, and books show that nursing continues to struggle for recognition as a profession in almost all parts of the world. Everywhere, nurses face difficulties in establishing the authority and value of their work. This is because most nurses are women, and "women's work" is still looked down on. Nurses are continuing to increase their authority and status by showing that their work is valuable and necessary, not just to patients but to society as a whole.

▲ A lecture on the use of bandages at a training school for nurses at the turn of the twentieth century.

A nineteenth-century woman brings medicine to a patient. Nursing was not recognized as a profession until schools were created to train qualified personnel well into the twentieth century.

NURSING ETHICS As nursing has developed as a profession, nursing ethics have developed along with it. At first, nursing emphasized hygiene, cleanliness, and kind caring for people who were sick or injured. In the late twentieth century, nursing emphasized the promotion of health, prevention of illness, restoration of health, and reduction of suffering. Nursing ethics have also changed. Nursing ethics used to involve rules of conduct for nurses to follow when they visited sick people, but nursing ethics has become a complex branch of bioethics.

Early Interpretations of Nursing Ethics

Early nursing ethics, developed by Florence Nightingale, involved the image of the nurse as a chaste, good woman in Christian service to others—an obedient, dutiful servant. This good nurse was committed to the ideal of doing what was right and saw nursing as a religious calling to serve others. She had high moral character, was disciplined by her moral training, and could be relied on to do her Christian duty and serve others.

Nursing as ministry. Early textbooks on nursing ethics were full of this image of the nurse as a good woman. In addition to being physically and morally strong, the good nurse was expected to be a dignified, courteous, well-educated, and quiet woman from a well-respected family. The nurse's work was seen as a ministry and religious service, done in the spirit of Jesus Christ. This emphasis on moral and religious virtue, duty, and service to others would be an important foundation for later thinking on nursing ethics.

Changing roles, evolving ethics. World War II changed the nurse's role in patient care. Before the war, nurses were expected to be a doctor's obedient helper. After the war, nurses began to become independent practitioners. They, not just the doctors, were responsible for action in a patient's care.

Nursing ethics evolved along with this change in the nurse's role. The nurse's responsibility was no longer simply obedience to authority or loyalty to the institution where she worked. Instead of carrying out the orders of others, nurses were now responsible for independent decisions in patient care, including ethical decisions.

The Development and Revision of Nursing Codes of Ethics

As professional nursing developed, nursing organizations discussed the need for a code of ethics to govern nursing practice. In the United States, nurses discussed a code of ethics for the first time at the 1897 meeting of the American Nurses' Association, but the Association did not accept a code of ethics until 1950.

The International Council of Nurses first adopted an international code of ethics for nurses in 1953 and revised it in 1965. A new code was adopted in 1973, and it was reaffirmed in 1989. Here are some of the code's most important statements:

Inherent in nursing is respect for life, dignity, and the rights of man.

The nurse, in providing care, promotes an environment in which the values, customs, and spiritual beliefs of the individual are respected.

This code was revised in 1960, 1968, and 1976, and it has been a model for nursing codes of ethics in other countries. In 1986, the code was revised again to describe ethical principles that guide nursing in the United States.

Many nurses' associations throughout the world have developed codes of ethics. They all agree on some things. Nurses are responsible for being competent and skilled in their work, they are expected to have good relations with coworkers, and they must have respect for the life and dignity of the patient. They must protect the patient's privacy, and they cannot discriminate against patients on the basis of race, religious beliefs, cultural practices, or whether the patient is rich or poor.

Moral Concepts of Nursing Ethics

Nurses must speak out for the rights and needs of the patients they serve. They must work with the patient and other caregivers to provide the best possible treatment, and they must care about the patient's well-being. These principles are important because they have been a foundation of nursing standards and ethical statements throughout the history of nursing. They also help define the relationship nurses have with their patients.

Advocacy. Advocacy is active support of an important cause. This means that nurses work to support the rights and needs of the patients they serve. Some people see this as a nurse's legal obligation. Others say advocacy means that nurses help patients take responsibility for their own health and their own lives. Still others believe that advocacy is a moral concept that defines how nurses view their responsibilities to patients.

Advocacy is said to be associated with courage and heroism, when the nurse must speak out on behalf of the patient. It can also be

▶ Nurses demonstrate for recognition in London (1974). Although nurses have proved their competence as health-care professionals, they are still considered with much less respect than doctors.

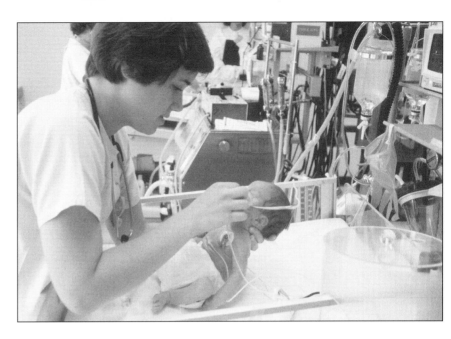

Even though nursing has gained recognition as a valuable profession, it is still thought of as a woman's job and attracts few men.

seen in quieter moments, such as when a nurse talks with a patient about what sickness, suffering, or dying means to the patient and his or her family. Two general ethical principles—respect for human dignity, and fidelity—are a part of advocacy. Some nurses describe advocacy as the ethical principle behind the work nurses do to protect the human dignity, privacy, choice, and well-being of the patient. This last view seems to match the values described in nursing codes of ethics and the major ethical responsibilities of the nurse.

Accountability. Nurses are accountable for their work. This means they are personally responsible for what they do, and they are expected to be able to give good reasons for their nursing decisions and actions. Nurses must know what their legal accountability is in order to be licensed. State boards of nursing also provide information on nurses' legal responsibilities in patient care. When nurses agree to care for patients, they also agree to be accountable for their actions, and to do their work according to the standards set by the state boards of nursing and others.

Accountability is a basic moral value and moral foundation for nursing. A few codes of nursing ethics have stated that accountability is a central moral concept of nursing. In the United Kingdom, a national nursing organization has provided a written description of how accountable nurses must be in professional practice.

Cooperation. Cooperation means that nurses work actively with others to provide good care for patients, to design nursing care, and that they do so in the same way that they want these professionals to work with them. Cooperation encourages nurses to work with others toward shared goals, and to sacrifice personal interests to maintain their relationship with other professionals. Cooperation has been included in many codes of nursing ethics as a moral concept of nursing practice.

> People feel better and heal faster when they know someone cares about them.

Caring. Caring is essential to the nurse's role and is believed to influence how people feel about their health and about life as a whole, both when they are healthy and when they are ill. Nurse caring is aimed at protecting the health and welfare of patients and shows a commitment to protecting people's dignity and their health.

Some thinkers believe that caring is related to the protection, welfare, and maintenance of another person. Others define caring as a moral obligation or duty among health professionals and say that caring and being involved with others includes being concerned about their experiences.

These views show two parts of the idea of caring. First, caring is a natural human response. Second, caring is related to social or moral ideals, such as the human need to be protected from cold or hunger and the human need for love. Caring can be seen as a special duty that people have for each other, and an ethical obligation for nurses.

Related Literature

Walt Whitman served as a nurse in the Civil War. His poem "The Wound Dresser" (1892) is one of several he wrote about the experience of nursing wounded men, many of whom died. Cleaning and dressing wounds as best he could, Whitman shows his compassion in his willingness to die for the young men he tries to help. His most effective aid may have been his personal caring, holding the dying men in his arms.

Michael Ondaatje's novel *The English Patient* (1993) focuses in part on a Canadian nurse who, at the end of World War II, refuses to abandon a dying patient whose body is badly burned. She nurses him in a bombed-out monastery as she and others try to uncover his identity. Her care includes reading to him and helping him to retrieve the memory of his life story.

Between the Heartbeats: Poetry and Prose by Nurses (1995), an anthology edited by Cortney Davis and Judy Schaefer, portrays many different aspects of nurses dealing with patients. Several themes emerge:

1. the devotion of nurses to their patients and to their profession;
2. the difficulty of not being able to cure some patients;
3. the necessary balance between objectivity and emotion; and
4. the special role of the nurse as an advocate of the patient.

PATIENTS' RESPONSIBILITIES AND RIGHTS

This entry consists of two articles explaining various aspects of this topic:

Duties of Patients
Patients' Rights

DUTIES OF PATIENTS It may sound strange to talk about duties of patients when all most patients want to do is concentrate on themselves and get better. Still, the effort to make someone better often takes many people—both health professionals and others—and the patient needs to behave in a way that allows others to do their job.

Duties Owed to Health Professionals

Two patient responsibilities are:

1. honoring commitments, including following a treatment plan the patient has agreed to. If a patient has received a prescription for a medicine that must be taken three times a day, the patient is responsible for taking the medicine on schedule.

2. providing information, especially if the information is necessary for the health-care provider to determine what is wrong with the patient and how to treat it. If a patient has blurred vision or a racing heartbeat as a result of taking illegal drugs, the patient is responsible for telling the doctor about taking the drugs.

No patient is required to agree to any treatment just because a doctor recommends it. Patients must be fully informed about the treatment and its benefits and risks, and then make an informed choice to follow the treatment or not. Once the patient agrees to a treatment, the patient should follow the plan. If the plan is impossible to follow, or the patient decides he or she does not want the treatment because it is inconvenient or because it has uncomfortable side effects, he or she must tell the doctor as soon as possible so that different treatment choices can be considered.

Patients do not have a right to demand treatments that have been proved not to work.

Duties Owed to Other Individuals

Duties patients owe to others are justified by the relationship between them and the patient. Patients have a duty to protect the interests of their families and to avoid hurting them. Patients may have a duty to tell them if they have a genetic disorder or an infectious disease. Patients might normally think this is private information, but they have a duty to tell family members in order to protect their health.

Unless citizens assume responsibilities—to serve their families, neighborhoods, and towns—life will be reduced to mere self-interest. If we want a right to trial by jury, we must be willing to perform jury duty. If we want a right to health care, we must assume the responsibility of living healthier lives.

If it is difficult to explain the relationship between the patient and the other person, it may also be difficult to know exactly what duties the patient owes. For instance, in the United States there is an ongoing debate about whether or not an unborn baby is a person in the eyes of the law, and what duties a pregnant woman has to protect the health of her unborn child. Should she stop smoking? Stop drinking? Does a woman have a right to abort a fetus, and if so, at what time during the development of the baby? Gaining a clearer idea of the nature of the fetus and the mother's relationship to it will help people make these decisions.

Duties Owed to Other Patients

A patient receives many benefits that would not be possible without the sacrifices made by patients in the past. You could not take medicine to cure an infection unless the medicine had been tested on volunteers in the past. You could not receive care from a highly qualified doctor or nurse unless that person had spent some time as a student, practicing on other patients, under the supervision of another doctor or nurse.

"It bothers me when people smoke three packs of cigarettes a day for thirty years and then stick me with their health bill."

—Jerry Brown
former governor of California.

Duties Owed to Society

Should your right to health care be based on how closely you have obeyed ideas of a healthy lifestyle? If someone exercises, does not smoke, and eats the right foods, should that person receive health care whenever it is needed? What about another person who has spent years drinking, smoking, and sitting on the couch, and who has become sick as a result? The second person is far more likely to become ill, and will require more care. Some health risks, such as smoking, are obvious, but it is difficult to determine exactly what a healthy, low-risk lifestyle is, because scientists keep discovering risks that no one knew about before.

Lifestyle. Some people argue that a person who does not live a healthy lifestyle should not receive care, because the illness is that person's fault. Others argue that a person who follows a high-risk lifestyle should pay a greater proportion of his or her health-care costs, or the person who follows a healthy lifestyle should be rewarded by having to pay less.

General Topics: Lifestyles

Addictions. Some of the debate about a person's duty to avoid health risks involves the fact that many risks, like smoking, alcohol use, and drug use, are addictive. Addiction usually means that the person has lost voluntary control over the drug use. If people cannot control what they are doing, it is useless to say that they have a duty to control their actions. You cannot have a duty to do what you are unable to do.

Most addicted people can control some parts of their behavior, even when other parts of the behavior seem impossible to control. Smokers may be able to sign up for "Quit smoking" groups or other treatments, but they may choose not to. Or, they may choose to go to events where others are smoking, and they know they will be greatly tempted to smoke.

Who Pays for Health Care?

Should smokers who get lung cancer as a result of smoking have to pay more for treatment than nonsmokers who also get lung cancer through no fault of their own? Or should smokers not receive treatment at all? To some extent, deciding whether a patient has the duty to remain healthy, or whether or not only some people are entitled to care, depends on how you value the individual's welfare in relation to the welfare of the community as a whole.

A purely individualistic view means that you have no responsibility to help pay for the health-care needs of anyone else. A purely communitarian view means that we all have a shared responsibility to provide decent care for everyone, and that this shared responsibility is weakened if we shut some people out of health care or charge them more because they have taken more risks with their health.

A duty to avoid health-care risks seems more justifiable when it is applied to everyone equally instead of being used to condemn some people whose choices and lifestyle might differ from others. One difficulty in applying this duty to everyone is that it is impossible always to know who is taking risks. How could one know if someone has really stopped smoking? Also, a duty to avoid health risks seems more justifiable if instead of condemning and spying on individuals, a more evenhanded kind of control could be used. Instead of refusing health care to people whose diseases are caused by smoking or heavy drinking, it might be better to put a tax on the sale of all tobacco and alcohol products.

Virtues of Patients

A virtue is a quality of a person that is generally admired or praised. In our culture, many people praise health-care givers, but few stop to think about the virtues of patients.

> Patients with liver disease have the responsibility not to use any alcohol for six months before they can be put on the waiting list for a liver transplant.

> On July 15, 1998, Senate Republicans proposed legislation to protect patients' rights. The plan seeks to give patients a greater choice of doctors and provide tax breaks to make insurance more affordable. But the Democrats' plan would cover 148 million people, while the Republican patients' protection provisions would generally affect only one-third that number, because they would apply only to self-insured employer-sponsored plans.

General Topics: Substance Abuse

In a book called *For the Patient's Good* (1988), Edmund Pellegrino and David Thomasma say that the virtues of a good patient include truthfulness, commitment to the healing plan and the healing relationship, justice, tolerance, and trust. These virtues are all related to the patient's obligations to the health-care provider and to other patients.

Courage. Courage is an important virtue for patients, who often wonder whether they have the strength to get through an illness or to do what is needed to cure or relieve it. Courage has two parts: accepting and enduring symptoms that cannot be cured or relieved, and fighting against the illness.

It is important for patients to be realistic. If people find out they have a lifelong condition, such as diabetes, the first thing they must learn to do is accept the fact that this is a reality. It will not go away. Then they have to act on this acceptance.

Hope. Hope is another virtue that is important for patients. Even when a patient is scared or depressed about an illness, hope can keep the depression from taking over.

Most people agree that certain virtues are especially important for patients. These include courage, wisdom, humor, hope, truthfulness, and faithfulness to the task of healing. Faithfulness to the work of healing may involve endurance through suffering, or through resistance to the disease.

Some of these virtues might seem obvious to most Americans, but in other cultures, where people have different values, patients might be expected to have very different virtues.

PATIENTS' RIGHTS In most industrialized countries, it is taken for granted that citizens have a right to medical care, but few people think about what rights people have once they are patients in a medical care system. In the United States, people may not have always believed that everyone has a right to medical care, but they have been very concerned about what rights people have once they begin getting that care. Only in the mid-1990s did people begin discussing whether or not all citizens had a right to medical care, or to medical insurance.

Origin of Patients' Rights

In the beginning, discussion of patients' rights focused on clinics and hospitals, where health care was provided, and emphasized issues from natural childbirth to informed consent. By the 1990s, patients' rights were beginning to be applied to all health-care settings and were also being discussed throughout the world.

Philosophers distinguish between positive and negative rights. Positive rights are rights to such necessities as food, education, health care, and housing. Negative rights basically come down to the right to be left alone—to practice one's religion, to speak freely, to join a political party. It is much harder for a society to establish positive rights because they cost a lot of money.

In February 1998, President Bill Clinton proposed a Bill of Rights for Patients who receive medical care through government programs. Patients would be entitled to appeal denials of insurance coverage, would be able to get information about doctors' and hospitals' performance, and would have easier access to specialists and emergency care.

Roughly one-third of all Americans receive medical care through the government:

1. federal employees
2. active military personnel and veterans
3. people on Medicaid and Medicare
4. Native Americans

The Switch from Trust to Money

Historically, the relationship between doctor and patient has been based more on trust than on money, unlike many other business transactions. This changed in the late twentieth century. Patients expected higher-quality care, and they were angered by increasing costs. They began seeing themselves as "consumers" shopping around for a good health-care deal. Hospitals started taking a more businesslike view of health care, and by the 1980s, many of them decided to operate as private businesses.

United States law has been on the side of the weaker person in disputes between someone in power and someone with less power, as in disputes between landlord and tenant, seller and buyer, creditor and debtor, employer and employee, police and suspect, and warden and prisoner. As hospitals became more powerful, the law began coming to the aid of patients asserting their rights in medical situations.

Basis of Patients' Rights

Recognition of patients' rights comes from two sources: (1) Health-care consumers have certain rights, which the consumers do not give up when they enter the health-care system; and (2) many doctors and hospitals do not recognize these rights, do not make sure they are protected or allowed, and often do not allow patients to exercise their rights or to fight for the recognition of their rights.

History

The Joint Commission on Accreditation of Hospitals is a private organization that sets standards for hospitals, and checks to make sure they meet the standards. In 1969, the Joint Commission asked for proposals to change and rewrite its standards. The National Welfare Rights Organization is an organization that works on behalf of consumers. In June 1970, this group wrote twenty-six demands that it wanted the Joint Commission to include in the standards. This list was the first complete listing of patient rights from the patients' point of view.

Patient Bill of Rights. In 1972, the American Hospital Association adopted a Patient Bill of Rights based on the idea that when a person receives care from a doctor in a big institution like a hospital, it is different from when a person receives care from a doctor who works alone or in a small clinic. The hospital, as an institution, is responsible for the patient, just as the doctor is. The Patient Bill of Rights includes the following rights:

1. respectful care
2. current medical information
3. information the patient needs in order to give informed consent to treatment
4. refusal of treatment
5. privacy
6. confidentiality
7. response to requests for service
8. information on other institutions related to the patient's care
9. ability to refuse to participate in research projects
10. continuity of care
11. examination and explanation of financial charges
12. knowledge of hospital regulations.

In 1992, two more rights were added: the right to see one's own medical records, and the right to tell health-care providers what care one wants to receive in the future. The list of rights is vague and is not complete, and there is no way to make sure that patients really have all these rights in every hospital, but it is a beginning. People need to know what rights they have as patients.

Between 1974 and 1988, thirteen states adopted a patients' bill of rights: Arizona, California, Illinois, Kentucky, Maryland, Massachusetts, Michigan, Minnesota, New Hampshire, New York, Pennsylvania, Rhode Island, and Vermont. By the late 1990s, all fifty states had adopted some form of advance health-care directive document. These documents allow people to tell others what kind of health care they will want if they become too ill to think clearly or to express what they need and want. Both former President Richard M. Nixon and Jacqueline Kennedy Onassis used these documents in 1994.

European movement. Americans may think and talk more about rights than people in other countries, but the United States is not the only country with a patients' rights movement. In 1975, the Parliamentary Assembly of the Council of Europe recommended that its sixteen member nations take action to ensure several things: (1) that sick people receive relief from their suffering and that people can prepare adequately for death; (2) that euthanasia, or mercy killing, be studied; (3) that doctors are told that sick people have a right to full information on their illness and its treatment; and (4) that sick people should be told about the routine, procedures, and special equipment of the hospital. By 1990, work on a European Declaration of the Rights of Patients had begun.

In 1992 Baby K was born without a higher brain and was not expected to live very long. The mother wanted her child to be treated aggressively for other ailments. The doctors refused because the treatments would not benefit the infant.

A worldwide trend. In 1991, a national conference on patients' rights was held in Japan, and patients' rights were beginning to be recognized there.

The worldwide trend toward recognizing human rights in health is part of a global movement for the recognition of human rights in general. The acknowledgment that a person has the right to determine what is done to his or her own body includes the right of a patient to refuse medical care.

Documents like the Nuremberg Code (1947), the United Nations Universal Declaration of Human Rights (1948), and the United Nations International Covenant on Civil and Political Rights (1966) are foundations of the patients' rights movement.

Mental Patients' Rights

The strength of a society's commitment to justice and humanity may be judged by looking at its treatment of people who are vulnerable, feared, or disliked. Mentally ill people have long been disliked, feared, persecuted, and set apart. A look at how these people are treated gives a useful perspective on issues in mental patient rights.

Colonial times. In the United States, mentally ill people have had many of their rights taken away. Since colonial times, the mentally ill rich have been treated differently from the mentally ill poor. Mentally ill rich people were usually kept at home, or in private institutions, and they were hidden from society to protect the reputation of their families. Poor people who were mentally ill have been left for the community to take care of.

If mentally ill people seemed harmless, their main threat was seen as using the community's money to keep them fed, clothed, and housed. To prevent this from happening, mentally ill people were often whipped and sent away from town, forced to wander from village to village. If they refused to leave their home town, they were often locked up in a jail or a poorhouse.

Nineteenth century. During the early nineteenth century, a Quaker minister, Thomas Scattergood, brought "moral treatment" of the mentally ill to America. Early doctors reported great improvement with his methods. In this treatment, mentally ill people were placed in asylums in the quiet countryside, where the doctor had absolute control. Life was very disciplined and included work by the patients.

This treatment was a big improvement in the treatment and conditions of mentally ill people. They still did not have the same legal and civil rights that other people had, but their treatment was humane and hopeful.

Twentieth century. This improvement did not last long. By the early twentieth century, the asylums were terribly overcrowded. Many new immigrants were arriving in America and the mentally ill population was growing along with the general population. Many of the patients stayed in asylums when they could not be cured.

> "I proceed, gentlemen, briefly to call your attention to the present state of insane persons within the Commonwealth, in cages, closets, cellars, stalls, pens—chained, naked, beaten with rods, and lashed into obedience."
>
> —Dorothea L. Dix (1802–1887), mental health reformer, in a memorandum to the Massachusetts legislature

Overcrowding and disorder became excuses for tying patients down and for harsh punishment. Asylums became warehouses for people instead of treatment centers.

Public concern about treatment of the mentally ill led Congress to establish the Joint Commission on Mental Illness and Health in 1955. The commission set a goal of mental-health care that meets the needs of all citizens. Community mental-health centers would provide mentally ill people with treatment close to their homes and jobs, and would reduce the need for long or repeated stays in hospitals.

The move away from mental institutions. In the 1960s and 1970s, drugs were developed that could affect a person's mood or thoughts. These drugs could help some mentally ill people to lead normal lives outside institutions. Also, community-based care expanded, people became more concerned about civil rights and more tolerant of mentally ill people living within the community, and the number of mentally ill people in hospitals dropped.

The process of moving mentally ill people from institutions to the community was not smooth. Many patients were sent out of the hospital without any help or training in getting along in the outside world. Most had no supervision for taking the drugs they needed. Often, people in the community did not like having patients around, since the patients were not prepared to live independently in the community.

Even though most people wanted improved care for the mentally ill, treatment in large state hospitals was still bad. Hospitals were too small, patients were poorly treated, and there was not enough staff to care for all the patients. During the 1960s and 1970s, these conditions led to many lawsuits, which brought more attention to the condition of mentally ill people.

These lawsuits involved three categories of patient rights: the right to treatment, the right to refuse treatment, and the right to the least restrictive treatment. In 1975, a case called *O'Connor* v. *Donaldson* related to a fourth right, the right to liberty. This right encompasses the other three.

The Right to Refuse Treatment

The right to refuse treatment includes almost all other rights of patients. It brings up questions about the amount of control that a health-care provider can have over a patient, especially if the patient does not want the treatment. The issues brought up by this right include the right to privacy, control over one's own life, the right to think whatever one wants to think, freedom from harm, freedom from cruel and unusual punishment, and the right to receive the least restrictive treatment available.

Many mentally ill people do not want to take drugs that may help them. This is because the drugs often have unpleasant side effects, which may include dry mouth, tiredness, and blurry vision.

The Declaration on the Rights of Mentally Retarded Persons was adopted by the General Assembly of the United Nations on December 20, 1971. An earlier text, adopted in 1968, had laid the foundation for a universally acceptable code. Following are some excerpts from the 1971 text.

The mentally retarded person has, to the maximum degree of feasibility, the same rights as other human beings.

The mentally retarded person has a right to protection from exploitation, abuse and degrading treatment.

Whenever mentally retarded persons are unable, because of the severity of their handicap, to exercise all their rights in a meaningful way or it should become necessary to restrict or deny some or all of these rights, the procedure used for that restriction or denial of rights must contain proper legal safeguards against every form of abuse.

In *O'Connor* v. *Donaldson* (1975) the Supreme Court was asked to rule in a case where a man, allegedly suffering from **delusions** (but described by all as harmless), was confined to a mental hospital for close to twenty-five years without receiving treatment. The Justices ruled that, even in the case of a mentally ill patient, the freedom to make certain kinds of decisions must be protected.

delusions a mental state characterized by false beliefs regarding the self or other persons or objects outside the self

Other side effects include sudden muscle movement that the person cannot control. This movement can be dangerous or embarrassing, or may make it difficult to do ordinary tasks.

Drugs or No Drugs

Some people refuse to take the drugs because they are used to the way their illness makes them feel, and they do not want to change it. The illness may make them feel important and in control, and it may make it easier for them to avoid problems that exist in the real world, including the illness itself.

In many states, and at the national level, mental-health lawyers have tried to expand and clarify the right to refuse treatment, especially when the treatment involves drugs that change a person's mood or thinking. The rights discussed include the patient's right to protect all thoughts, feelings, and beliefs from government interference, the right to protect one's own body, and the right to choice of treatment.

The Courts Protect Patients' Rights

The courts that decided these cases wanted to set in place methods to protect patients' independence and power to refuse to take treatment drugs. These courts tried to give all patients the right to refuse drug treatment, but some people say this is not practical. For example, the courts assumed that mentally ill patients were able to decide what was best for them. But because the patients are mentally ill, they may not be able to decide what they really need and want. They may not be able to act in their own best interest.

This battle for and against patients' rights continues. On one side is concern for patients' ability to make their own decisions and to be protected from potential danger. On the other side is concern for patients' medical needs and the fact that some patients may need treatments that they do not want in order to get better. Future discussions of this right will have to include not only legal and medical views of mental illness but also the effect that court intervention may have on patients and society.

Right to the Least Restrictive Alternative to Hospitalization

In 1974, in the case of *Dixon* v. *Weinberger*, a judge ruled that patients in the District of Columbia have a right to be treated in the least restrictive alternative to being placed in a hospital. The District of Columbia and the federal government were responsible for identifying and transferring patients to new community hospitals.

By 1994, the court's orders had still not been completely followed. This case shows that the courts are limited when they try to

establish rights. If a community does not want to obey and provide the necessary services, little can be done to enforce the court order.

Another important court case that is related to the least restrictive treatment right is *O'Connor* v. *Donaldson*. In this case, the court said that states have an interest in providing care and assistance to patients, but it also said the patient's preferences should be recognized. Just because patients are mentally ill does not mean that they do not know whether or not they want to be at home or in a hospital. Although the state governments can order a person to be placed in a hospital, most people who are able to live at home should be allowed to stay there.

The court's emphasis on a patient's ability to survive at home, and the expectation that the least drastic treatment right to be used, have put more pressure on communities to provide more services for mentally ill people. But many communities may take a long time to get the money, workers, and other resources they need before they can provide services.

The Mentally Ill in Society

To answer the question "What rights do mental patients have?" it is important to go beyond court cases, government actions, or laws, and look at the status mentally ill people have in American society. Historically, the rights of mentally ill people have been ignored and denied. People thought mentally ill people were unable to make their own decisions, and society's concern was to place them in institutions where they could not hurt themselves or others, and where they could be treated.

The patients' rights movement, which includes civil rights lawyers, enlightened mental-health caregivers, and former patients, has struggled in courts, in legislatures, and in local communities to stop patient abuse, end negative stereotypes, increase community services, and give patients the power to claim their full civil rights. Major struggles in the areas of right to treatment, right to refuse treatment, right to least restrictive alternatives, and right to liberty have led more people to recognize these rights. But it is clear that more work is needed to prevent past abuses from happening again.

see also

Ethics and Law: Information Disclosure, Truth-Telling. and Informed Consent

PATIENT RESPONSIBILITIES
American Medical Association
1993

The American Medical Association's (AMA) Patient Responsibilities draws upon the recognition that successful medical care depends upon a collaborative effort between doctors and patients. Originally published in July 1993, Patient Responsibilities specifies responsibilities of patients for their own health care.

The background section of the original report states: "Like patients' rights, patients' responsibilities are derived from the principle of autonomy.... With that exercise of self-governance and free choice comes a number of responsibilities." Here is a summary of the report.

1. Good communication is essential to a successful physician-patient relationship. To the extent possible, patients have a responsibility to be truthful.

2. Patients have a responsibility to provide a complete medical history, to the extent possible, [on all] matters relating to present health.

3. Patients have a responsibility to request information or clarification about their health status or treatment when they do not fully understand what has been described.

4. Once patients and physicians agree upon the goals of therapy, patients have a responsibility to cooperate with the treatment plan. Patients also have a responsibility [. . .] to indicate when they would like to reconsider the treatment plan.

5. Patients should discuss end-of-life decisions with their physicians and make their wishes known. Such a discussion might also include writing an advance directive.

6. Patients should be committed to health maintenance through health-enhancing behavior. Illness can often be prevented by a healthy lifestyle, and patients must take personal responsibility when they are able to avert the development of disease.

7. Patients should also have an active interest in the effects of their conduct on others and refrain from behavior that unreasonably places the health of others at risk.

PROFESSIONAL–PATIENT RELATIONSHIP

People who write about the ethical issues in the relationship between health professionals and patients have written mostly about a doctor and a patient. However, equally important ethical values, principles, and virtues apply to the relationship between patients and the other health professionals.

Moral Models of the Relationship

Deciding what is right, good, and moral between health-care professionals and patients depends on one's moral point of view. There are several very different moral models of this relationship, as explained by different philosophers and ethicists.

The ethicist Robert Veatch has offered four models of the doctor–patient relationship. The "engineering model" views the doctor as a scientist dealing only with facts, not involved with moral questions. The "priestly model" views the doctor as someone with greater knowledge and status than the patient, and assumes that the doctor has the right to make moral and value-laden choices for the patient. The "collegial model" assumes that the doctor and the patient are equals, like friends, with equal trust, loyalty, and roles. The "contractual model" involves a mutual understanding of the benefits and responsibilities that each person has.

Nursing perspective. Sheri Smith has offered three models from a nursing perspective. In the "surrogate mother" model, the nurse is morally bound to take complete responsibility for the well-being and care of a passive patient. In the "technician" model, the nurse is responsible only for the technical, medically skilled parts of the patient's care. In the "contracted clinician" model, the nurse's responsibility is determined by the values and rights of the patient. This model also assumes that patients are able to decide what is best for them.

covenant a formal, binding agreement; a pledge between two or more parties

Contractual model. Many people view the contractual model as the best. Some say that the most important part of it is the fact that patients have the right to make informed decisions about what the best care is for them. Others say the key to this model is that it emphasizes patients' best interests. But some people do not believe this model is the best possible model.

Covenant. William F. May believes that the idea of "covenant," instead of "contract," provides the best moral model. In a covenant model, the health-care professional not only provides service to the patient but also shows gratitude to the patient and society for being able to provide the service. This idea of gratitude pushes the professional to go beyond the bare minimum of simply providing services, as in a contractually based relationship. In addition, the idea of a covenant reminds the professional to go beyond competent treatment of an individual patient, and extends to society as a whole.

Paul Ramsey also proposes a covenant model in his book *The Patient as Person: Explorations in Medical Ethics* (1970). He believes that in a covenant model, the patient and professional clearly owe each other certain moral attitudes and actions. He emphasizes loyalty. Informed consent, which means that the patient is told all the risks and benefits of a treatment and then makes an informed choice of whether to accept the treatment or not, is a primary aspect of loyalty in the professional–patient relationship. Ramsey stretches loyalty beyond the idea of a simple contract between the patient and the professional, and says that the patient and the professional must acknowledge full partnership and loyalty to each other as ethical human beings.

Ethical Principles in the Professional–Patient Relationship

Ethical principles are important in the professional–patient relationship. They provide insight into its ethical foundations. Some of the most important are respect for persons, the rule of doing no harm, beneficence, truthfulness, autonomy, and justice.

Respect for persons. Respect for persons, which means showing that you know that each person has dignity and worth, is a part of most professional codes of ethics, mission statements of health-care organizations, and patient-rights documents. It is also found in many other writings on ethics. This principle assumes that people have inherent worth and deserve respect simply because they are human beings. Different philosophical, religious, and scientific discussions of this idea have provided a wide basis for this concept in the health professions.

This principle is a challenge to health professionals. One part of this challenge is to define what is meant by "a person." The other

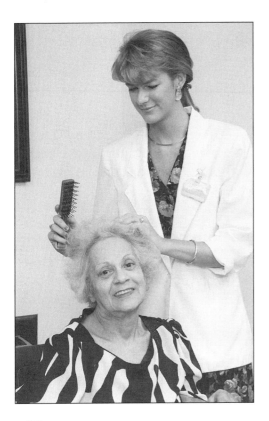

A caring nurse brushes the hair of an elderly patient.

Judeo-Christian historically based in both Christianity and Judaism

challenge is to use the idea that all persons deserve respect as a guide for behavior. The answers to these difficult questions may seem obvious at first, but they are not. Two different health-care professionals may agree that all people are valuable and have rights simply because they are persons. They may base this idea on the **Judeo-Christian** belief that all people have value because they are children of God, or they may base it on the philosophical idea of Immanuel Kant, that "persons should be treated as ends and not as means to ends." But how do these health-care providers decide whether or not to perform an abortion? The decision is difficult even if the professionals believe in the worth of each person, because it is difficult to define whether or not the unborn fetus is "a person."

Even though it sometimes leads to difficulties, this principle is an important part of the relationship because it asks health professionals always to show respect for patients as persons.

The rule of doing no harm. This rule is often called the first principle of medical practice. No one knows when this idea began in health care, but its meaning and usefulness are clear.

Patients go to doctors and other health professionals with the expectation that that they will receive whatever appropriate care the doctor is equipped to provide. They do not expect careless, negligent, or other harmful care.

The duty to do no harm also makes the professional aware that society as a whole, not just the patient, expects that he or she will not cause harm. Discussions of harm must take into account that sometimes it is necessary to cause pain or other harm in order to heal or help the patient. For example, in surgery the patient is cut open with a knife, but this is done to remove or change something that is causing pain or an illness.

Beneficence. The principle of beneficence is doing things for the good of other people. It is very important in understanding the professional–patient relationship. Since the earliest times, people have gone to healers expecting that the healer's first duty is to help the patient. People still believe that the health-care professional has the moral duty to work for a patient's well-being, guided by that person's health-related concerns and needs. Any other goal, such as learning more about a disease and its cure, or earning money, or making sure the hospital or clinic is receiving enough business, is not considered an appropriate guide for the professional's treatment of the patient.

Ethics and Law: Beneficence

Combined with the principle of respect for persons, the principle of beneficence is the idea that health professionals have a moral obligation to do good for everyone who needs their help. They should be willing to treat all kinds of patients. This idea is tested when the doctor is prejudiced against people of a different race, people from a particular ethnic group, old people, gay people, or any other group, and, because of prejudice, when the doctor finds

it hard to treat everyone with equal respect. A health professional may also call a patient "undesirable," "difficult," "disgusting," or "hateful" if the patient is dirty, drug-addicted, or irritating. In every case, the professional must see the patient as worthy of treatment, no matter how different the patient's life and values are from the professional's. If the differences are so great, and the professional is unable to overcome prejudice and provide good care, then the professional must ensure that the patient receives good care from someone else.

Truth-telling. Philosophers often treat the principle of truth-telling as a separate principle, but some say that being truthful comes from principles of respect for other persons. Because truth-telling is usually considered to be always good and always right, some situations can cause an ethical dilemma for professionals. Health professionals believe that patients want doctors to help them keep up hope when they are in bad shape. Therefore, doctors may lie about a patient's true condition. If a patient is paralyzed and there is a great chance that she may never walk again, should the doctor tell her? What if this causes her to get so depressed that she is unable to get better in other ways? If there is even the smallest chance that she can walk again but the doctor says it is unlikely, she may get so depressed that she loses the chance for recovery and does not even try to walk again, even though she might have, if she had kept up hope. On the other hand, it can be dishonest for a doctor to keep up a patient's hope when there is very little reason to expect that the patient will get better.

So, for centuries, doctors have sometimes protected patients from the truth. This was thought of as a "benevolent lie," and the doctor took responsibility for taking advantage of the patient's belief that the doctor would always tell the truth. The "good" effect of the lie, which was that it helped the patient to have hope, was thought to outweigh the fact that it is still a lie.

In the latter part of the twentieth century, this idea about not telling patients the complete truth began to change. Most people have come to believe that if patients know the complete truth, they are better able to take control of their lives, and this is better for the patient. In other words, hope does not depend only on whether or not the patient knows the truth but also on the doctor's honesty and respect for the patient's ability to deal with illness and make choices about how to react to it.

Autonomy and self-determination. Patient autonomy means that the patient is in control of his or her life, and makes decisions about it. It also means that patients should be given the facts about their illness and its treatment, so they can make informed choices about it.

The idea of the patient's autonomy has changed over the years. Originally it was thought of as the patient's *choice* to refuse treatment. This expanded to the idea that the patient had a *right* to

In an article published in 1927, Dr. Joseph Collins stated: "The longer I practice medicine the more I am convinced that every physician should cultivate lying as a fine art."

A 1961 study indicated that 90 percent of all doctors would not tell their patients the truth if they had cancer.

Ethics and Law: Information Disclosure, Truth-Telling, and Informed Consent

Ethics and Law: Autonomy

refuse treatment. It has evolved into the idea that a patient has the right to play a central role in determining the course of treatment. This makes the patient much more powerful in the professional–patient relationship. In 1990, the U.S. Congress passed the Patient Self-Determination Act, which made it a law that patients be able to tell doctors what treatments they want in critical situations.

Conflict sometimes develops between the patient's right to choose and the professional's obligation to do what is best for the patient. Sometimes patients do not want treatment that would actually be good for them. Should the doctor overstep a patient's wishes and force him or her to accept treatment? How does the doctor know when the patient wants something that is not helpful, wants something that is actually harmful, or does not want something that is helpful? And what does the doctor do in these situations? These questions must be resolved as they occur. They are always difficult for both patients and doctors.

Justice. The principle of justice holds that each person should get what is due to him or her. The nurse or doctor and the patient can expect several things to happen. First, the patient can expect to be treated fairly. The rules of health care will be fair, and other people will not be given any advantages. In a waiting room, people will be accepted in the order of first come, first served, even if one of them is the doctor's friend. The only people who can morally justify "skipping the line" are those who are in greater danger and need treatment sooner.

The principle of justice also raises ethical questions related to scarce resources. Health professionals have a duty to help patients, but this duty is somewhat flexible. Some treatments are scarce and very expensive, so that not everyone who needs them can have them. Two patients may both need a liver transplant. One is young and otherwise healthy, and would live a long life with a new liver. The other is old and has many other health problems, and even with a transplant would probably not live a healthy life. There are few livers available for transplant, and doctors may choose to provide the transplant for the young and healthy person. This may not seem completely just, but sometimes scarce resources force difficult decisions.

The health professional who is committed to justice must work to preserve justice within the professional–patient relationship and also must try and use resources fairly. If patients do not receive justice in health care because of discrimination against their race, ethnic group, religion, gender, age, or other factors, they do not receive just treatment. Health professionals must work against these social barriers to provide good care for everyone.

Compassion and considerations of caring. Compassion has long been included in discussions of virtues that health professionals should have. The word *compassion* comes from Latin roots meaning "to suffer with." There are different theories about what this

see also

Ethics and Law: Justice

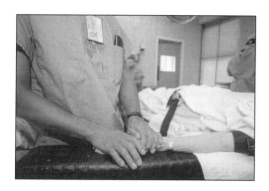

 A nurse holds a woman's hand as she lies on the operating table.

see also

Ethics and Law: Care and Compassion

might mean in health care. One central idea is that patients heal better when professionals show that they understand and sympathize with the patient's situation. This means that professionals not only care about whether or not they have done their duty but also are sensitive about patients' feelings. The moral question for professionals is not simply a question of telling the truth to the patient but also of helping the patient to understand and live with what the truth really means. A doctor may have the duty to tell the truth and tell a patient that a disease cannot be cured, but then the doctor also has the responsibility to help the patient adjust to this painful and difficult knowledge. The professional should sympathize not only with the patient's obvious suffering but also with the patient's deeper losses. Even if the patient is not physically in pain, it is a virtue to have compassion for the patient's deeper losses, which may be psychological or spiritual.

The idea of "caring" is considered a major value in the professional–patient relationship. This idea shows how professionals might show compassion in their everyday work. Making caring the most important part of the professional–patient relationship may be a key part of healing when human suffering and deep loss are involved.

Mechanisms for Resolving Ethical Conflict in the Professional–Patient Relationship

Ethical issues in the professional–patient relationship have received new attention in everyday health care. Inevitably, professionals and patients often have very different opinions about what is right. This conflict does not always mean that the two sides are angry at each other. More often, it is a sign that both professionals and patients are frustrated and are having trouble deciding what is the ethical and right thing to do.

Patient representatives. There are several ways for patients to find help when they disagree with a professional about their care. First, most hospitals have a patient representative, who helps patients and their families when disagreements come up. The patient may tell the representative that he or she is not receiving good care. The patient representative may help the patient talk to the doctor about this, so it can be resolved, or may refer the patient to someone else who can help.

Ethics consultants. Second, some major hospitals are hiring ethics consultants. These people deal with ethical questions involving patients and their care. The ethics consultant may work with the patient, the patient's family, the doctor, the nurse, or other professionals. Usually the consultant meets with everyone who is involved, and helps them to understand the ethical questions that are a part of the problem, talks about the issues, and recommends ways to choose between the different ethical issues. The consultant does not make the final decision. It is made by the professional and patient together.

Ethics committees. Third, clinical ethics committees are part of many health-care systems. They usually include a mix of doctors, nurses, and other health-care professionals, and their work is similar to that of the ethics consultants. Sometimes an ethics consultant is called first, and if this person thinks the problem would be solved better by a group of people with different experience and opinions, the ethics consultant may call the ethics committee to meet and work on the problem.

Everyone agrees that in health care it is better to prevent a moral or ethical problem than to try and resolve it later. The professional must communicate clearly with the patient and with other professionals, must be skilled and competent, and must care about the welfare of the patient. These are keys to preventing ethical conflict, and they are also powerful tools for resolving ethical conflict when it comes up in the professional–patient relationship.

Ethics committees are found primarily in facilities with more than 200 beds (less than 1 percent of U.S. hospitals in 1995) and in hospitals offering teaching programs. The major benefits of ethics committees include decision making, providing legal protection to the medical and hospital staff, developing policy on life support, and serving as a forum for the resolution of professional disagreements.

Related Literature

Probably the story used most frequently to raise issues about professional–patient relationships is William Carlos Williams's "Use of Force" (1938). Narrated by a physician who makes a house call, this story shows him losing his temper as the sick, frightened child refuses to open her mouth so he can look at her throat. Believing that the child has diphtheria and that he must make a diagnosis immediately, he brutally forces her mouth open in a clear abuse of power.

Richard Selzer, a physician-writer, has written several collections of stories and memoirs about physician–patient relationships. One story, "Brute," from *Letters to a Young Doctor* (1982), describes an emergency-room physician who has great difficulty controlling a belligerent, drunk "brute" of a man who has to get stitches. Finally, the doctor sutures the man's ear lobes to the operating table in order to make him hold still.

PROFESSIONAL ETHICS

expertise advanced skill

Among a society's most important institutions are the social structures by which it controls the use of specialized knowledge and skills. How can a society control the use of important, specialized **expertise** and enable those outside the expert group to reap the benefits and enjoy the values that depend on it? One of the most important social structures developed to gaining this end is the institution of the profession.

The Key Features of a Profession

For an occupation to be a profession, it must provide its clients with something that the larger community deems extremely valuable. Health and the preservation of life are valued by almost everyone. This means that we value those in the health professions—doctors, nurses, and so on—for keeping people healthy and for saving lives.

Expertise. The expertise of a profession has elements that are very complex. We should rely only on those people who have been educated and trained in these skills by experts in the field to do the work of that profession.

In addition, only those who are educated in both background and practice of a profession can correctly judge the need for experts in a given situation. For this reason, only experts can judge the quality of other experts' work.

Institutional recognition. The expertise of a profession is not only recognized by those who possess such specialized knowledge. It is also recognized by the members of the larger community. Because of the exclusive nature of professional expertise and the importance of applying it, outside recognition often takes the form of certificates or licenses that the larger community grants to members of a profession. In this way, community members have formal authority over the profession's expertise.

Because clients value the activity of a profession, and because only experts, not ordinary people, can judge the performance of other experts, clients usually grant to a profession's members independence in the actual practice of the profession.

Professional Obligations

One of the most important features of a profession is that its members have to accept a set of performance standards. To make this point clear, compare and contrast a "standard" picture of a profession with a "commercial" one.

According to the commercial picture, practicing a profession is no different from selling goods in the marketplace. Those who take this view feel that the professional has a product to sell and makes the needed agreements with interested buyers. Beyond some basic duty not to pressure or cheat others, professionals would have no other obligations than performing what they said they would. A person would have no special duties just because he or she is a professional.

However, when a person enters a profession, he or she does undertake obligations that are continually being agreed upon or adjusted by the professional group and the larger community. In the standard picture, there are duties over and above those in other relationships. Professionals have obligations, and the content of these obligations for each profession is its "professional ethics."

The word "profession," as used today, describes just about any type of work that requires some technical knowledge and training (e.g., auto mechanic, beautician). Historically, professions also contained a moral element of commitment and self-sacrifice for the benefit of others. Examples of the latter were teachers, doctors, nurses, and lawyers. Even members of these professions often lose sight of higher ideals and are interested in career advancement only.

Nineteenth-century caduceus. The caduceus has become the symbol of the medical profession.

There are eight basic categories of questions about professional obligations:

1. Who are this profession's chief clients?
2. What are the main values of this profession?
3. What is the ideal relationship between a member of this profession and the client?
4. What sacrifices do members of this profession have to make, and how do the duties of this profession take priority over its members' other moral considerations?
5. What makes someone competent in this profession?
6. What is the ideal relationship between the members of this profession and other professionals?
7. What is the ideal relationship between the members of this profession and the larger community?
8. What must the members of this profession do to keep their commitment to its values and to educate others about them?

As is true for many other human institutions, if we did not have the institution of the profession, we would need to invent it or something like it in order to live together effectively. Obviously, we live in a world where no one person can have all the knowledge and skills on which the achievement of so many of life's values depend. But, like other human institutions, each profession needs to be looked at regularly to make sure it continues to fulfill its purpose. One of the main roles of bioethics is to provide the members of the health professions and the larger community with the tools to use to make sure that these professions are performing in a way that is ethical and right.

The American Medical Association (AMA) was founded in 1847. Its first tasks were to establish standards for medical education and to formulate a code of ethics. The code of 1847 was adopted by both the AMA and the New York Academy of Medicine.

Related Literature

Ernest Hemingway's story "Indian Camp" (1925) raises many ethical issues. The physician-father has been called from his fishing trip to an Indian camp because a woman is having trouble delivering a baby. The doctor takes his young son along with him, and never has him leave the scene in spite of its gruesome nature. Making racist comments about Indians, the white doctor performs a C-section on the woman with his pen knife, using no anesthesia and sewing her up with fishing line. The woman's husband, unable to stand his wife's agony and the arrogant racism of the doctor, slits his own throat.

David Hilfiker, author of Healing the Wounds: *A Physician Looks at his Work* (1985), left family practice in a small town in Minnesota to practice "poverty medicine" among men dying of AIDS

Related Literature, cont'd

in Washington, D.C. (described in his second book, *Not All of Us Are Saints*, 1994). In his first work, Hilfiker describes medical mistakes and the problems that doctors have in admitting that they make errors, even though such mistakes often have major effects on the health and well-being of patients and their families. One of his worst mistakes was believing tests that said a woman was not pregnant and then accidentally aborting her living baby, thinking that it was a dead fetus.

◆◆◆

Anne Sexton's poem "Doctors" (1975) opens with a list of the many things doctors do competently, such as diagnosing disease, prescribing drugs, operating, and in other ways trying to repair damages to the human body. But they cannot always save lives; they cannot always heal. She says that the doctors "are not Gods/ though they would like to be." And though most doctors abide by the Hippocratic oath to do no harm, some are too arrogant, acting sure of themselves until life finds a way to humble them.

TRIAGE

Triage is the method by which medical personnel—doctors and nurses—assess which patients will be treated first, second, and so on, according to patient need. When medical personnel, operating rooms, beds, drugs, and other resources are limited, and immediate treatment of all patients is not possible, patients are "sorted." Such sorting enables doctors and nurses to use the available supplies in the most effective way. Triage is a process of screening patients on the basis of their immediate medical needs and on the probable medical success in treating them.

It has become common to use the term "triage" in cases where medical personnel need to make decisions about assigning scarce medical resources in general. But unlike the everyday practice of allotting medical resources, triage usually takes place during crises. At these times, medical personnel need to make quick decisions about the critical care of many patients. Triage enables doctors and nurses to establish priority for treatment. Usually, these decisions are based on the utilitarian idea of doing the greatest good for the greatest number of people.

History

The history of triage dates back to the early nineteenth century. Baron Dominique Jean Larrey probably organized the first plan for classifying military **casualties**. Larrey was Napoleon's chief medical

casualty a person or thing that has been lost, injured, or destroyed

officer, and he insisted that those who were most seriously wounded be treated first, regardless of their military rank.

Military Use

Larrey's plan for sorting battle casualties influenced later military medicine. But the practice of systematically sorting casualties did not become common until World War I. It was also at this time, between the years 1914 and 1918, that the term "triage" became part of military medicine in Great Britain and the United States. The term (from the French verb *trier*, meaning "to sort") originally referred to sorting agricultural products such as wool and coffee. In military medicine, "triage" was first used both for the casualty treatment itself and for the place where such screening happened. At the *poste de triage* (casualty clearing station), soldiers' wounds were assessed for their seriousness and for the need to evacuate the soldiers quickly to hospitals in the rear of the field.

The following triage categories have become standard, though terms may vary:

1. *Minimal.* People whose injuries are slight and need little or no medical care.
2. *Delayed.* Those whose wounds, such as burns or bone fractures, need medical attention that can be delayed for some time without increasing the chances of death or disability.
3. *Immediate.* People whose injuries, such as severe bleeding or blockage of breathing passages, require immediate treatment.
4. *Expectant.* Those whose injuries are so great that there is little or no hope of survival, given the available medical resources.

▶ Color-coded triage tags showing level of urgency.

▶ A Red Cross nurse ministers to severely wounded soldiers. The massive casualties suffered during World War I forced the military to create triage, a sorting procedure to allow delivery of care to those who needed it most urgently.

First priority is for those soldiers in the "immediate" group. Next, as time and resources permit, treatment goes to the "delayed" group. Other than providing comfort, little medical attention is given to those in the "expectant" category. Those in the "minimal" group are left to take care of themselves until all other patients are cared for.

Saving the greatest number of lives. From the beginning, according to William W. Keen in his book *The Treatment of War Wounds* (1917), the obvious reason for such sorting of wounded soldiers was based on this idea: "The greatest good of the greatest number must be the rule." The good to be achieved was saving the greatest number of soldiers' lives.

But the rule could also mean doing the greatest good for the military effort. When interpreted in this way, triage could produce very different priorities. The least injured could be treated first so that they could be returned quickly to battle.

Maintaining military strength. As military triage developed during the twentieth century, the goal of maintaining fighting strength did, in fact, become the greatest goal. As Gilbert W. Beebe and Michael E. DeBakey wrote in their book *Battle Casualties: Incidence, Mortality, and Logistic Considerations* (1952), "Traditionally, the military value of surgery" is "not merely a matter of saving life; it is primarily one of returning the wounded to duty, and the earlier the better."

Civilian Use

The nuclear weapons used in 1945, at the end of World War II, introduced a destructive power never before seen. In the nuclear age, triage plans have had to include the possibility of countless hopelessly injured civilians. In earlier days, it was not unusual to plan for one or two thousand casualties from a single battle. Now, triage planners must consider the probability that one nuclear weapon

could produce a hundred times as many victims or more.

Triage has moved from military into civilian medicine in two important areas: the care of disaster victims and the operation of hospital emergency departments.

The need for triage in hospital emergency departments is due, in part, to the fact that a number of patients needing immediate emergency care may arrive at the same time. They could temporarily overwhelm the hospital's emergency resources. More often, however, the need for triage in hospital emergency rooms comes from the majority of patients waiting for routine, nonemergency care. Emergency-department triage is often taken care of by specially trained nurses who use elaborate ranking methods to determine the seriousness of injuries or illnesses.

Ethical Issues

Traditionally, it is the duty of doctors and nurses to protect the interests of patients as individuals and to treat people equally on the basis of their medical needs. These same commitments to treating patients equally have, at times, been used for the treatment of war casualties. The Geneva Conventions—international agreements establishing the care and protection in wartime of the sick or wounded—call for medical treatment of all casualties of war solely on the basis of medical needs. In triage, however, health-care professionals think of patients in groups and give priority to goals such as maintaining military strength. Triage puts aside loyalty to the individual patient in order to accomplish the most good or prevent the most harm. The tension between remaining faithful to the individual patient and the goal of seeking the greatest good for the greatest number of people is triage's most important ethical issue.

Many Americans first learned about triage from the 1970s TV series M*A*S*H*. The opening scene of each episode showed helicopters arriving with wounded soldiers at a field hospital and doctors quickly determining which of the wounded needed attention first.

Victims of the first atomic bomb. In August 1945, Hiroshima, Japan, was hit by the first atomic weapon ever used. The resulting demand for medical attention by those who survived the attack prompted officials around the world to devise triage plans for medical emergencies.

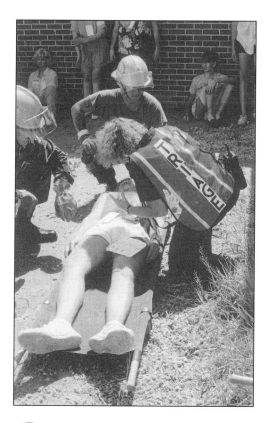

A triage officer tends to an accident victim.

Triage also generates a number of other questions concerning what is right or ethical. To what extent are the utilitarian goals of military or disaster triage appropriate in assigning everyday medical care, such as beds in an intensive care unit? If some casualties of war or disaster are considered hopeless, what care, if any, should these patients receive? Should their care include mercy killing?

Triage is a permanent feature of present-day medical care in military, disaster, and emergency situations. As medical research continues to produce new and expensive therapies, it will be tempting for the health-care professions to use the rules of triage for making decisions about who gets what care. This is true for other issues such as world hunger and population control. But the wisdom of using the lessons of medical triage for such social ills is doubtful. It will be up to "the experts" to carefully question triage's application to social problems.

Index to Volume 2